Microphones for the Recording Musician

MUSICIAN'S GUIDE TO
home recording

Microphones for the Recording Musician

Phil O'Keefe

Craig Anderton

Hal Leonard Books
An Imprint of Hal Leonard LLC

Published in 2018 by Hal Leonard Books
An Imprint of Hal Leonard LLC
7777 West Bluemound Road
Milwaukee, WI 53213

Trade Book Division Editorial Offices
33 Plymouth St., Montclair, NJ 07042

All photos of miking techniques are provided courtesy of Phil O'Keefe.

Book design by NextPoint Training, Inc.

Library of Congress Cataloging-in-Publication Data

Names: O'Keefe, Phil (Recording engineer) author. | Anderton, Craig.
Title: Microphones for the recording musician / Phil O'Keefe, Craig Anderton.
Description: Montclair, NJ : Hal Leonard Books, 2018. | Series: Musician's
  guide to home recording | Includes bibliographical references.
Identifiers: LCCN 2018043864 | ISBN 9781540035639
Subjects: LCSH: Sound recordings--Production and direction. | Popular
  music--Production and direction. | Microphone.
Classification: LCC ML3790 .O45 2018 | DDC 621.382/84--dc23
LC record available at https://lccn.loc.gov/2018043864

www.halleonardbooks.com

Contents

Acknowledgments

From Phil O'Keefe: There are many people I want to thank for making this book possible.

The recording engineers who have taught, mentored, and inspired me over the years—Bruce Swedien, George Massenburg, Bill Dooley, Phil Ramone, Roger Nichols, Geoff Emerick, and many others. Hopefully this book will in some small way pay it forward for all that I've learned.

Every musician who ever let me put a microphone in front of them, as well as all of the musicians who participate on the Harmony Central forums. I've learned so much from all of you.

My Mom and Dad for their love, support, and help; my daughters Angele, Kelsey, Samantha, and Anastasia for their inspiration; my wife Sandy for her love and encouragement and for putting up with the long hours I've spent in the studio or in front of a computer writing; and finally, a special thanks to my one and only Aunt Mo for being my biggest fan and for always encouraging me.

Most of all, I'd like to thank you—the reader. My sincere hope is that you'll find this book useful, and that it helps you to capture some great music you can share with a world that is always in need of more of it.

From Craig Anderton: A series like this is never the work of one person, but a collection of experiences obtained over the years from too many people to acknowledge here. Yet some deserve a special mention.

Dan Earley, my editor at Music Sales, who was the first person to say, "You know what would be cool? A series of books on recording, like those Time Life libraries." Well Dan, better late than never, right?

The team at Hal Leonard—especially John Cerullo, who green-lighted this series and brought in Frank D. Cook to serve as the editor for these books.

My father, who taught me that it didn't matter if I was a dreadful writer as long as I could edit my words into something readable—and who also showed me what it meant to love music.

My mother, who with my father was unfailingly supportive when I wanted to do things like drop out of college, join a rock band, go on tour, and never look back!

My brother, who understood music on a very deep level and died too young.

And of course, the many *(many)* engineers and producers who let me look over their shoulders and absorb knowledge like a sponge over the past five decades. My hope is that this series will help pass their collective wisdom on to another generation.

Introduction

About This Book

Welcome to the book series Musician's Guide to Home Recording. This series of short publications was written to address the needs of musicians and recording enthusiasts who are interested in creating self-produced songs or doing audio production work for others.

Rather than trying to cover all aspects of recording in a single sprawling volume, each title in the series concisely and accessibly addresses a particular subject. You can select individual titles to hone in on certain skills or proceed through the entire series; this kind of approach lets you develop a comprehensive knowledge at your own pace.

This book, *Microphones for the Recording Musician*, covers microphone types and polar patterns; mic accessories, connection options, and preamp considerations; microphone setup and stereo miking techniques; and much more.

The Importance of the Microphone

Microphones (also called *mics* or *mikes*) capture the sonic "raw materials" that we sculpt into our finished aural masterpieces. Unless you're working only with samples and virtual instruments, microphones are essential—so you'll need to decide how many you need, which mic technologies will best serve your purposes, and the specific models that will hit the sweet spot of performance, budget, and suitability. This book is here to help you make the right decisions.

Fortunately, microphone-manufacturing techniques have improved dramatically over the last few decades. This has led to lower prices as well as higher quality. Home recording enthusiasts can afford to have a variety of mics in their collection; however, it is more important than ever to choose the right mics for your particular needs rather than compiling a random selection of mics just because you can.

Some mic designs are intended for specific applications, while others are more general purpose. For example, the mic built into your cellphone plays a very different role than the mic an international spy might use to bug a room, or the multi-thousand dollar mic a professional engineer might use to record a symphony orchestra. There are also different mic technologies that may be better suited to recording some instruments than others.

What's more, mic accessory improvements have kept pace with microphones themselves. Today's mic preamps, which are needed in most cases to amplify the mic's low level, have noise and distortion performance that rivals the very best microphone preamps of not long ago. Even the ones built into relatively inexpensive audio interfaces will do a more than credible job.

Ultimately, though, what's even more important than the technology itself is the expertise needed to apply that technology. This book explains why moving a mic just a centimeter or two can make a huge difference in terms of sound quality, whether you're miking a guitar amp or a drum kit. Years of experience miking a variety of projects—jazz, punk, classical, acoustic, rock, and more—have been distilled down to these pages. Each chapter ends with a Key Takeaways section that summarizes the most important points in the chapter. For those who want to dig deeper, I've included "Tech Talk" sidebars with additional information.

Ready to make some music? Don't click record just yet… there's a lot of information in here that will help you select and set up your mics to make better recordings.

Tips and References

This book includes various tips, definitions, cross-references, and other supplemental nuggets throughout its pages. These are denoted with the following icons and formatting.

 Tips and side notes provide helpful hints and suggestions, background information, or additional details on a concept or topic.

 Definitions provide explanations of technical terms, industry jargon, or abbreviations.

 Cross-references alert you to another section, book, or online resource that provides additional information on the current topic.

Chapter 1

Microphone Basics

The most common mics convert acoustical energy (the movement of air molecules in a sound wave) into electrical energy. Once transformed into electrical energy, we can amplify, transmit, or record the signal. Devices that convert one form of energy into another are called *transducers*. Other transducers include loudspeakers that convert electrical energy into acoustical energy, magnetic guitar pickups that convert string vibrations into electrical signals, and LEDs that convert electrical energy into light.

Microphone Types and Technologies

The first mics were developed in the 1870s. Although they were noisy and inefficient, the technology evolved rapidly. The three most common mic technologies for today's musicians and recording engineers are condenser mics (invented in 1916), moving-coil dynamic mics (invented in the early- to mid-1920s), and ribbon mics (also invented in early- to mid-1920s).

You don't need to know exactly how mics work, as long as you know which part of the mic the sound waves should hit and which part plugs into your audio interface, preamp, or mixer. However, each type of microphone does have different audible characteristics, so your ears will tell you when to use one type over another.

 For detailed information on the technology behind different types of microphones, see Chapter 4 of this book.

Condenser Mics

Condenser mics are very common in the studio, but less so live. Prices are generally reasonable, even for a high-quality condenser mic, because manufacturing techniques developed over the last few decades have made it easier to produce high-quality mics at ever-lower prices.

Power Requirements

Condenser mics need power (typically 48 volts), which can come from one of three sources:

- ◆ An internal battery

◆ Phantom power—Mixers, mic preamps, and other devices designed to accommodate condenser mics can send power through the mic's audio cable to a condenser mic.

 Phantom power gets its name from the fact that you don't see any physical power connections to the microphone.

◆ External power supply—Mics that use a tube need more voltage than phantom power can provide. These use a power supply, sometimes called a *brick* because—well, it looks like a small brick.

 We cover powering microphones in more detail in Chapter 3 of this book.

Condenser Mic Pros and Cons

Here are some of the pros and cons to consider with regard to condenser mics:

◆ Condenser mics excel at capturing fast *transients* (a sound's initial sound waves).

◆ The sonic quality is generally bright and open (the sound isn't boxy or muffled).

◆ They generally have a wider frequency response range than other mic technologies.

◆ High-quality, high-end condenser mics can be expensive.

◆ Condenser mics aren't too common in live performance because they're somewhat sensitive to physical damage if hit or dropped, and are sensitive to moisture.

◆ Most condenser mics need to mount on stands because they're larger and more unwieldy than dynamic mics.

Diaphragm Size

Diaphragm size is an important specification for condenser mics.

◆ Compared to *large-diaphragm* condenser mics, *small-diaphragm* mics are less sensitive to low levels, can handle higher levels, and are more sensitive to high frequencies. They're excellent mics for acoustic percussion instruments. This application also takes advantage of the condenser mic's inherent fast transient response.

◆ Large-diaphragm condenser mics are more sensitive to low levels and are less sensitive to high frequencies. They're good for vocals (except for "screamers") and for recording softer sounds like room ambience, nylon-string guitars, ukuleles, and environmental sounds.

Dynamic Mics

Dynamic mics are the most popular mic technology for live performance, but they are also commonly used for recording. Some singers even prefer how their vocals sound with dynamic mics. Unlike condenser mics, dynamic mics don't need a power source.

Dynamic Mic Pros and Cons

Here are some of the pros and cons to consider with regard to dynamic mics:

- Dynamic mics tend to be fairly inexpensive.

- They're the most rugged mic type and are relatively immune to physical shock, wind, and moisture.

- Because they're well-suited to live performance, dynamic mic bodies are usually designed to be hand-held. This may be important for singers who like to hold their mic in the studio.

- Dynamic mics have the poorest high frequency and transient response of the three main mic technologies.

Diaphragm Size

Like condenser mics, dynamic mics offer large- and small-diaphragm varieties, and the diaphragm sizes exhibit the same general characteristics as they do for condenser mics. Small-diaphragm condenser mics are a common choice for guitar amps.

Ribbon Mics

Ribbon mic technology is a variation on that of the dynamic mic. Ribbon mics fell out of favor for a while, but the technology is having a resurgence due to improved manufacturing techniques, increased ruggedness, and lower costs.

Ribbon Mic Pros and Cons

Here are some of the pros and cons to consider with regard to ribbon mics:

- Ribbon mics have excellent transient response.

- Recording engineers often describe ribbon mics as sounding warm, detailed, and natural. These characteristics have made them popular in this age of digital recording.

- The mic's polar pattern makes ribbon mics a good choice for miking dual guitar cabinets.

 Microphone polar patterns are discussed in the next section of this chapter.

- Many ribbons have limited high-frequency response. This may require adding a bit of high-frequency EQ to the sound to compensate.

- Most ribbons have very low output levels and require a low-noise mic preamp with plenty of gain (60-70 dB).

- Ribbon mics tend to be expensive.

- Ribbons mics are inherently fragile; a windblast can destroy the ribbon element (especially with older models; some modern ribbon mics are more hardy).

Microphone Polar Patterns

Many analogies can be made between photography and recording. No camera lens covers all photographic needs. Some lenses are wide-angle types, others can focus in tightly, and some even provide special effects.

Microphones are similar: no single microphone is ideal for every situation. Some are designed to pick up sounds coming from any direction. Others capture sound that arrives from specific directions, while rejecting or attenuating sounds from other directions. A mic's *polar pattern* defines how it reacts to sounds coming from different directions.

Additionally, some mics significantly *color* sound that arrives *off-axis*, meaning they alter the tone of sounds from outside of the intended pickup pattern. Furthermore, stereo-miking techniques may combine two or more mics—sometimes with completely different polar patterns—to capture the stereo sound field. For these reasons, a good basic understanding of polar patterns is crucial to get the most out of your mics.

Understanding Polar Pattern Plots

Manufacturers often publish microphone polar patterns as a two-dimensional diagram that attempts to show a three-dimensional, 360-degree field surrounding the mic. The plot places the microphone at the center, pointing at the 0° mark on the graph (Fig. 1.1). This is the "on-axis" point for the microphone, with 90° and 270° being directly to either side of it, and the 180° mark directly to its rear. A plot line on the graph shows the microphone's sensitivity to sound arriving from different directions.

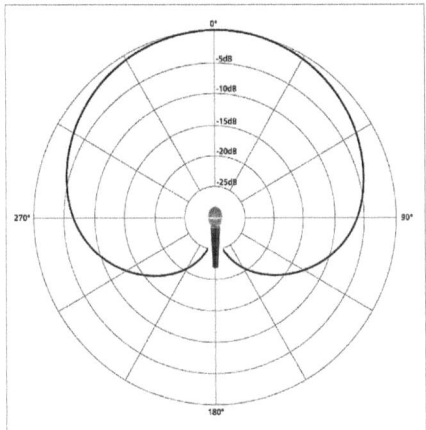

Figure 1.1 A typical polar pattern plot. This mic is designed to pick up sounds in front of it. Note the decreased sensitivity to sounds arriving at 180°—the mic's rear.

The plot line that's relative to the polar plot's concentric circles indicates the microphone's sensitivity to sounds arriving from that angle. A plot line on the outer edge of the outer circle indicates no attenuation. Each ring closer to the center represents 5 dB of attenuation. The closer the plotted line is to the center of the graph, the less sensitive the microphone is to sounds arriving from that angle.

The pattern in Fig. 1.1 is a *cardioid* (literally heart-shaped) pattern, and is most sensitive to sound arriving at 0°, on-axis, or from directly in front of the microphone. It's least sensitive to sounds arriving from 180°, or directly to the mic's rear.

You can test this yourself. If you have a Shure SM58 or similar cardioid microphone, sing into the center of the ball from about 6 inches away. Then flip the mic upside down and try to sing directly into the other end, with your mouth next to the cable's XLR connector. The distance from your mouth to the mic element is about the same in both cases, but the volume will be drastically different due to the cardioid pattern attenuating sound arriving from the rear.

Because polar patterns tend to vary at different frequencies, a single graph will often show more than one plot, with the response at different frequencies represented by different types of lines—for example different colored lines, solid lines, dashed lines, or dots and dashes.

The "Big Three" Polar Patterns

Although other patterns exist, the majority of mics use *omnidirectional, cardioid,* or *figure-8* polar patterns. Many mics have a single, fixed pattern, while some microphones offer interchangeable, removable capsules with different polar patterns. Still others have multiple, user-selectable patterns—usually selected via a small switch on the mic's body. (Some tube mics include a knob or switch on their remote power supply for this purpose.) Let's take a closer look at each of the three main polar patterns.

Omnidirectional Microphones

An omnidirectional microphone is designed to respond equally to sound arriving from any angle and at any frequency (Fig. 1.2). Omnidirectional mics are ideal for picking up the sound of multiple performers or group vocals. They usually have the most balanced sound and *flattest* frequency response (capturing sound evenly from low to high frequencies) because off-axis sounds aren't colored.

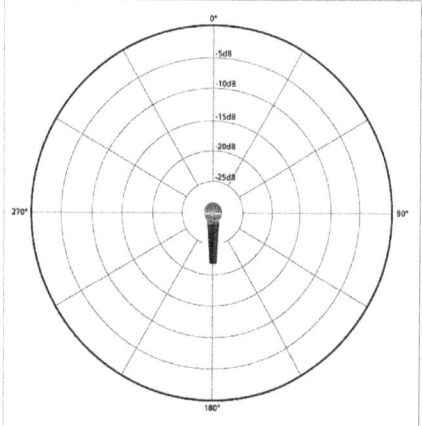

Figure 1.2 Omnidirectional microphones capture sounds arriving from any direction.

Omnidirectional mics don't exhibit the *proximity effect* (a phenomenon in which the bass response increases as the sound source gets closer to the mic), so they're excellent for close-miking without sounding boomy. This also means there's no loss of bass response as the mic moves farther away from the sound source.

For a natural sound, an omnidirectional mic usually can't be beat. This makes them very popular for classical and jazz recording. You can adjust the ratio of direct sound (from whatever you're recording) to the reflections from a room's surfaces by moving the mic closer to or farther away from the performers. Omnis (as they're often called) also tend to capture more of the sound of the room's character than other polar patterns. Therefore, they work best in rooms with good acoustic qualities.

Because they pick up sound arriving from all angles equally well, they don't reject feedback well. This makes them less suited for live sound. They may also have difficulty isolating individual instruments when multiple sound sources are in the same room, resulting in mic bleed. Regardless, in the studio or for live recordings, they are excellent for capturing natural, uncolored sound. They are often the mic of choice in A-B stereo or *spaced-pair* configurations (described in Chapter 5, along with other stereo miking techniques).

Bleed is sound from other instruments or sources besides the one you're trying to capture.

Cardioid Directional Microphones

Most microphones feature a cardioid polar pattern, as seen in Figure 1.1. Cardioid mics are most sensitive to sound coming directly from the front and less sensitive to sound arriving at the mic's rear. This makes them excellent for live performance, because if you point the mic toward the sound source and the rear toward stage monitors and PA speakers, the natural attenuation helps prevent feedback. This is a classic example of using a microphone's polar pattern to your advantage.

Sounds arriving off-axis are subject to coloration as well as attenuation. This can be an issue with cardioid mics, especially when recording multiple sound sources in the same room. Increasing the distance between the microphones can help reduce this.

A cardioid mic's unique polar pattern can be equally useful in the studio. For example, by placing two musicians so that they're facing each other, but on opposite sides of the room, and then aiming a cardioid mic directly at each, it's often possible to record more than one person at a time in the same room while still retaining relatively good isolation. Cardioid microphones are common in stereo miking setups, including the XY stereo and ORTF stereo configurations described in Chapter 5.

Bi-Directional or "Figure-8" Microphones

Microphones with figure-8 (also called bi-directional) polar patterns (Fig. 1.3) are less common than omnidirectional and cardioid mics, but they are still extremely useful.

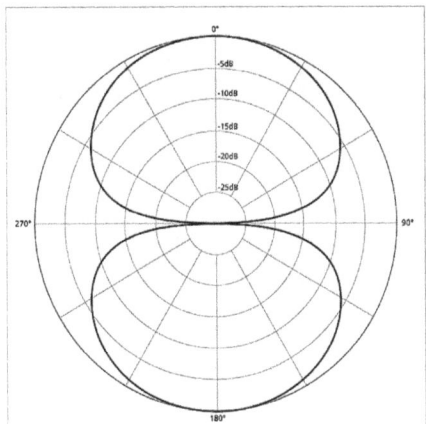

Figure 1.3 A figure-8 polar pattern rejects sound arriving from the sides.

Most ribbon microphones have bi-directional polar patterns, as do many high-end multi-pattern, large-diaphragm condenser microphones. Their significant proximity effect means that placing them close to a sound source will boost the low frequencies. You can use this to your advantage to make thin-sounding singers and instruments sound *fatter* (with more bass and depth).

Bi-directional mics have the best rejection at the null points of any polar pattern—many ribbon mics have nearly perfect rejection of sounds arriving from the sides. This makes them a great choice when recording two sound sources in close proximity, while still maintaining enough isolation to record them on separate tracks. For example, bi-directional mics are excellent for recording a single person performing on acoustic guitar and doing vocals simultaneously. To do so, aim the vocal mic's front at the vocalist's mouth, while pointing the side toward the acoustic guitar. Meanwhile, aim the guitar mic's front at the guitar, and the side (where rejection is near-perfect) toward the singer's head.

Bi-directional microphones are also essential for many stereo mic configurations, including the Blumlein Stereo Pair technique and mid-side stereo configurations described in Chapter 5.

 If you're getting too much hi-hat bleed in the snare mic, try using a figure-8 pattern mic on the snare and angle it so that the hi-hat is to the snare mic's side, with the snare directly in front of it. You'll get lots of snare on the recording, and less hi-hat bleeding into the snare mic than with other mic types.

Other Polar Patterns

There are several other polar patterns, but most of them are variations on the three primary types.

Subcardioid

Also known as a *wide-cardioid* pattern, the pattern is a cross between an omni and a cardioid, with less rear rejection. It has more sensitivity to sound arriving from the sides than a standard cardioid microphone, and somewhat more directionality than an omni mic.

Supercardioid

This is also similar to a cardioid pattern but with increased directionality, and some increased sensitivity to sound arriving from the rear (but not quite as much as the hypercardioid pattern—see next.) Typically, the best rejection is for sound arriving at 150° and 210°—not quite from right behind the mic, but close to it.

Hypercardioid

Like the quite similar supercardioid pattern, this is more directional than a cardioid, with even greater sensitivity to sound arriving from the rear than the supercardioid. This pattern's directionality is somewhere between a cardioid and a figure-8 pattern (Fig. 1.4).

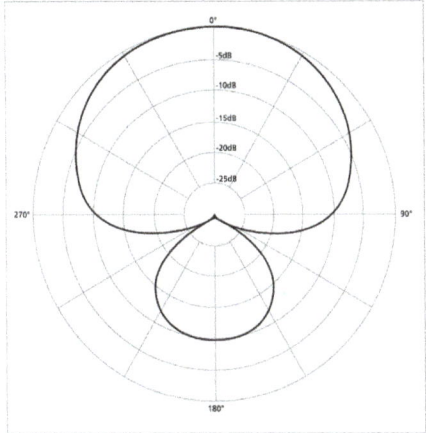

Figure 1.4 Note the similarities and differences between a hypercardioid polar plot and the cardioid or figure-8 polar patterns. Best rejection is at 120° and 240° off-axis.

Sometimes *More* Is More

Having a selection of different polar patterns can be very useful, because they can provide the tools needed for a variety of stereo recording techniques described later in this book. Large-diaphragm condenser mics with multiple, user-selectable polar patterns are useful because one mic can serve different purposes. Some small-diaphragm condensers are available with alternative screw-on capsules (Fig. 1.5); different polar patterns expand their utility as well.

Figure 1.5 Some small-diaphragm condensers like this Oktava MC-012 have exchangeable capsules so you can change the mic's polar response.

Rejection — It's All in the Angles

Where you aim the mic's front isn't all that matters: where you aim the null points to prevent sound pickup can also make a big difference in your recordings. So pull out those manuals, or fire up the computer to research your mics, and study their polar patterns. Then start experimenting with different mic placement angles on your recordings, with an ear toward not only what you want to capture, but what you want to reject.

Key Takeaways

♦ Condenser mics need power. They're excellent for capturing fast transients and have an open sound with a wide, high-frequency response. But they are more delicate (and generally more expensive) than dynamic mics.

♦ Dynamic mics tend to be fairly inexpensive and rugged, but they have the poorest high-frequency and transient response of the three main mic technologies.

♦ Dynamic and condenser mics have different diaphragm sizes. Smaller diaphragms are less sensitive to low levels, can handle high levels, and are more sensitive to high frequencies.

♦ Larger diaphragms are more sensitive to low levels and are less sensitive to high frequencies.

♦ Ribbon microphones have superior transient response. But they have a little less high-frequency response than condenser mics, require lots of preamplification, and tend to be expensive.

♦ A microphone's polar pattern shows where the mic is most likely and least likely to pick up sound. Some mics are more directional than others, so sometimes mic placement is about what a mic rejects as well as what it picks up.

♦ Directional microphones exhibit the proximity effect, where moving the mic closer to a sound source increases the bass response.

♦ Omdirectional mics that pick up sound from all directions are free from the proximity effect.

Chapter 2

Microphone Accessories

Microphones don't exist in a vacuum. Some accessories are internal to a mic, while others are external to it.

High-Pass and Low-Pass Filters

High-pass and low-pass filters are represented with the "bent line" graphical legend on microphones, mixing boards, mic preamps, and EQ plug-ins. The associated switches or controls apply *filters*, which are a type of equalizer that can change the signal's frequency response.

A symbol like the one in Fig. 2.1 identifies a switch or button for a high-pass filter (often abbreviated as HPF). This type of filter reduces low frequencies while passing high frequencies.

Figure 2.1 An HPF graphic shows that the filter reduces the low frequencies.

The symbol graphically represents what happens to the signal when it passes through the filter. As with a frequency plot graphic, such as the one shown in Fig. 2.2, the low frequencies are always on the left side of the X-axis, while the high frequencies are on the right.

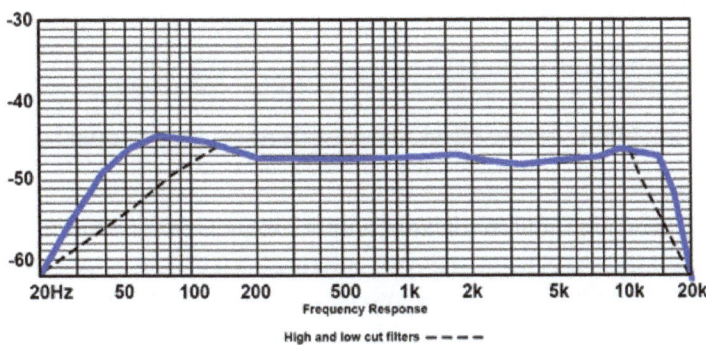

Figure 2.2 A typical frequency response graph. This one is for the Lauten Audio 320 microphone.

Note the dotted line toward the left in Figure 2.2 above. This indicates the frequency response with the high-pass filter enabled, which attenuates the low frequencies.

Some mics also have low-pass filters. These allow the low frequencies to pass through, while rolling off the high frequencies. A low-pass filter indicator indicates rolloff at the high end of the frequency spectrum, as shown with a symbol like Fig. 2.3.

Figure 2.3 Low-pass filter graphic

Note in Figure 2.2 that this mic also includes a low-pass filter. The dotted line toward the right shows how enabling the low-pass filter attenuates high frequencies.

Filter Cutoff and Slope

Some microphones, such as the Schoeps CMIT 5 U (Fig. 2.4), have multiple settings for high-pass filtering. While the actual frequency response itself may not be shown, the switch graphics will often give a general idea of what the different settings do.

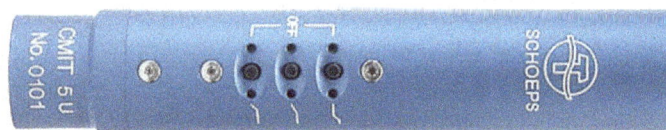

Figure 2.4   The Schoeps CMIT 5 U has a high-pass filter with different slopes. The switch on the left has a steeper cutoff slope than the switch in the middle.

A high-pass or low-pass filter has two main settings, although they're not always user-adjustable:

♦ The *cutoff frequency* is where the filtering begins to take effect. At this frequency, the response has already dropped by −3 dB.

♦ The *slope*, measured in decibels (dB) per octave, indicates how rapidly the response rolls off with respect to the cutoff frequency. For example, engaging a high-pass filter with a 6 dB per octave slope at 80 Hz will attenuate the signal by 6 dB at 40 Hz (an octave lower), and another 6 dB at 20 Hz (two octaves below 80 Hz) for 12 dB of total attenuation at 20 Hz.

A 6 dB per octave slope is considered fairly gradual. Steeper slopes, like 12 dB and 18 dB per octave, are also common. These roll off the signal past the cutoff frequency more dramatically than a 6 dB per octave filter.

When to Use Filters

When recording sounds that contain mostly midrange and high frequencies, high-pass mic filters can help produce a cleaner track by reducing extraneous low frequency noises (e.g., semi-distant machinery, room rumble, air conditioning, sounds from distant traffic, etc.). Full-range sound sources that emphasize low frequencies (e.g., upright bass, electric bass amps, and kick drums) are rarely recorded with a high-pass filter.

Low-pass filters de-emphasize high frequencies, and are much less common. Most mixing boards, hardware equalizers, and mics lack low-pass filters (the Lauten Audio Series Black LA-320 mic is a notable exception), although they're common in equalizer plug-ins. A low-pass filter may reduce snare and cymbal bleed when miking a kick drum, attenuate hiss from a noisy piece of gear, or tame overly bright vocalists, narrators, or instruments.

How to Use Filters

Engaging a filter is simple: just push the button, click the switch, or click on the virtual button and listen. Experimentation is key with low-pass and high-pass filters. Do you like what you hear? Only you can decide if a particular filter is right for the track and musical context.

Pad Button

This button reduces the microphone's internal level so that it can handle louder signals without distorting. You would typically engage this when positioning a mic right up next to a guitar amp, snare drum, or other loud sound source.

Acoustical Screens and Shields

Acoustical shields minimize room reflections from entering back into your mic after bouncing off walls. Shields became popular for use with vocal microphones first (Fig. 2.5), but models are now available for use with instruments and amplifiers, such as the sE Electronics Reflexion Filter. The tradeoff is potential coloration due to the proximity effect, because no sound-absorbing material is 100% absorptive.

Figure 2.5 The Primacoustic VoxGuard VU blocks room reflections from getting back into the mic. It's designed specifically for vocalists; note the window so the singer can see the engineer, or other band members.

With vocals, you might think that singing close to a mic makes any sound reflections from a room's walls, floor, and ceiling irrelevant, but your room influences the vocal's sound. Although you may be able to use the room sound to good effect, in general you don't want the room sound to have too much influence—the more natural the vocal, the more easily you can tweak its sound when mixing. Ways to isolate the mic from room noise include singing in a linen closet (all that soft material dampens the sound), using a screen that blocks reflections from the walls, or spending the bucks for a vocal booth, like the enclosures available from WhisperRoom.

Protective Case or Pouch

Good microphones are a significant investment and can be somewhat fragile, so treat them with care. You'll want to cover up a studio microphone or store it in a case when not in use. Covers and cases come in a variety of different styles, from inexpensive vinyl and cloth bags bundled with some mics to foam-lined, touring-type cases with die-cut cutouts for multiple microphones. There are also deluxe, velvet-lined hardwood cases suitable for your most hallowed studio microphone. All of them will help protect your mics from scratches, and some will also protect them from drops (Fig. 2.6).

Figure 2.6 The pouch on the left comes with Audio-Technica's AT2020USB mic, which is a USB mic designed for musicians on the go. The box on the right protects Audio-Technica's AT3035 condenser mic in the studio.

Protection is particularly important for condenser microphones because their electrically charged diaphragms tend to attract dust. However, cases are also good insurance for ribbon microphones and even dynamic microphones.

 If a mic is set up for a session and you don't want to remove it from its stand, cover it with a large, inexpensive plastic food storage bag with a desiccant bag taped inside.

Pop Filters

When you sing close to any kind of mic, you create bursts of air from plosives (*b*, *p*, and similar sounds). These can overload the mic and produce unpleasant, low-frequency popping sounds. There are several types of pop filters made for reducing these sounds. Fur-covered zeppelins are available to protect a mic in high-wind conditions outdoors (you might have seen these in on-location newscasts), while foam windscreens have been used on stage and in the studio for decades. Some mics include internal pop filters within their housing.

The most common pop filter for studios places a fine mesh (metal or nylon) between the vocalist and the mic to help diffuse bursts of air. Although some engineers feel pop filters detract from a vocal, pops can detract from a vocal even more. If you don't need a pop filter, that's probably because you're singing quite a distance from the mic, or are very careful about where you aim your voice when you're singing. When mixing, you can use equalization to reduce pops, but a pop filter minimizes the problem at the source. Engaging a mic's low-frequency rolloff switch can also help reduce pops.

Pop filters range from really cheap (an old nylon stocking stretched on a hanger in front of the mic) to really expensive. Most pop filters run between $20 and $100, and they'll each do the job to one degree or another. On the high end of the scale, there's the $300 PaulyTon Pauly SuperScreen pop filter (Fig. 2.7), which is exceptionally effective.

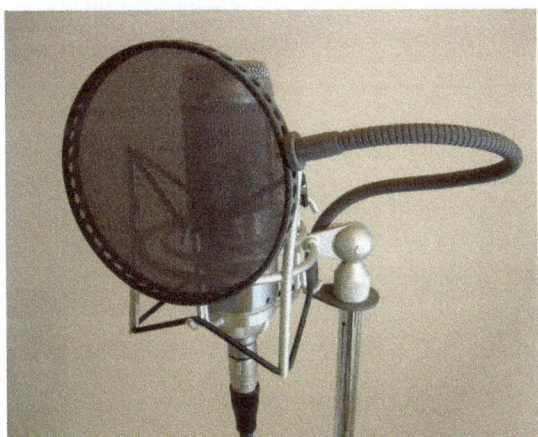

Figure 2.7 The PaulyTon Pauly SuperScreen is expensive, but its ability to stop pops is impressive.

Microphone Cables

For recording purposes, you'll need to connect your mic to an audio interface or other preamp, which requires a suitable cable. As with so many other aspects of audio, people argue for hours about cables. Some may opine that only a $200/foot cable made from Venusian unobtainium is acceptable for recording, while others routinely use the cheapest cables they can find. Cables can make a difference, but the cable length and its placement in the studio may have more impact than anything else. In a typical small studio, where cable runs are short and not draped over items like transformers that generate electric fields, you don't need significantly expensive cables. On the other hand, if you have rare German mics and the world's finest preamps, you may be able to hear a subtle contrast with expensive cables. Realistically, most people won't hear a difference.

Professional mics terminate in a male XLR balanced connector; avoid mics with other types of connections. XLR is a type of connector, while balanced wiring is a protocol that minimizes hum and noise pickup when carrying signals. With most mics, the only cable you'll need to record your vocals is one that has a female XLR connector at one end for plugging into the mic and a male XLR connector at the other end for plugging into the audio interface or preamp (Fig. 2.8).

Figure 2.8 A microphone cable has XLR connectors at each end. The connector on the left is male, the one on the right is female.

Tech Talk: How Balanced Lines Work

Balanced wiring uses two signal conductors plus a ground. One conductor line carries the original signal and the other carries a phase-inverted copy of the original. As an audio signal's voltage increases along one line, its mirror image on the other line decreases. This type of signal has to feed a special type of input, found in transformers and certain types of amplifier designs, that responds only to the difference between these two signals. The input circuitry flips the inverted signal back in-phase and sums it with the original signal to produce the final output.

This differential input rejects signals that spill over into both signal lines equally, such as hum and noise. Because the differential input responds only to the differences between the two signal lines, when the same signal is present on both lines, it becomes phase-inverted on the input and cancels itself out.

Tube microphones typically use multi-pin connectors and multi-wire cables to connect the microphone (which houses the tube electronics) with the microphone's power supply box. This box powers the microphone and houses the XLR output jack for connecting to an external mic preamp or mixing board.

Most tube mics come with the necessary cables and accessories, so you won't need to buy them separately. However, if you're buying used, make sure the tube microphone includes all necessary cables.

Tech Talk: Are USB Mics Any Good?

In addition to standard mics with XLR outputs, USB mics are available that can plug directly into your computer. These have some advantages: cables are less expensive, they don't carry analog audio, and you don't need a traditional audio interface. USB mics got a bad reputation because the earliest models were designed for convenience and low price, so performance was nowhere near a professional level. However, times have changed, and many USB mics today qualify for professional uses such as podcasting. Nonetheless, devices that connect with computers have a disturbing habit of becoming obsolete, whereas XLR mics will continue to be usable for the foreseeable future. If you need to connect an XLR mic to a USB input for mobile recording on a laptop, you can use an XLR to USB converter such as Shure's X2u (Fig. 2.9).

Figure 2.9 Several XLR-to-USB converters are available, but avoid the ones with rock-bottom prices—while good for karaoke, don't expect high quality from the preamps and converters.

Mic Stand

Not everyone needs a mic stand—some vocalists don't use them live, and even when recording, some singers will hold the vocal microphone. But those are exceptions. For the rest of us, and for miking musical instruments, you'll need a good stand to go along with your mic (Fig. 2.10).

Figure 2.10 This K&M mic stand has a boom attached at the top to make it easy to position the mic when miking various instruments.

Dozens of different mic stand types are available: desktop models, short stands, straight and boom stands, decorative ones for live shows, new high-tech models like the Triad-Orbit stands, and heavy-duty stands that can hold the largest and heaviest studio microphones (such as those made by Latch Lake). DynaMount even offers motorized, remote-controlled stands that let you change mic placement in the studio from the comfort of the control room.

 Clamp-on stand attachments can mount a second mic (such as for a performer who is singing and playing an acoustic guitar at the same time).

Microphone Shock Mount

The shock mount isolates the mic from a mic stand, which is important if there are other sounds in the room. Because a mic stand sits on the floor, any vibrations from the floor (from footsteps, a drummer's feet on a kick drum or hi-hat pedal, or a bass amp's output) can transfer up the stand to the mic, and add low-frequency artifacts. While these vibrations and thumps might not seem like much at the time, they can add up. In some cases, they may even ruin an otherwise excellent recording. A high-pass filter switch can help, but it's best to play it safe and isolate the mic as much as possible with a shock mount. Mics often include this mount as part of the package (Fig. 2.11).

Figure 2.11 The shock mount on the left is included with the Neat Microphones Worker Bee, which is inserted in the mount. The shock mount on the right is for an Audio-Technica AT3035.

Many high-end studio microphones come with suspension-style shock mounts as a value-added accessory. Similar mounts are available as after-market accessories for other mics. Compared to clip mounts and ring mounts that fasten mics to their stands, shock mounts are a bit more complicated. Most use an elastic band suspension system that cradles the mic and isolates it from physical shocks. While it takes some time and effort to insert the mic into the shock mount (and to remove it when you're tearing down), the advantages outweigh the extra effort.

Check with your mic's manufacturer to see if there's a purpose-built model specifically for your mic. If so, this will often be your best option. If such a unit isn't available, universal shock mounts that work with a variety of mics are widely available and relatively inexpensive. Adding a few to your collection and using them regularly will make a significant improvement to your recordings' sound quality.

Mic Preamps

Just as an electric guitar without an amplifier is only part of an instrument, a microphone also needs a good preamp. Most audio interfaces include built-in mic preamps, but more exotic stand-alone preamps (with tube amplifiers, audio transformers, and other high-end features) are also available. The right mic/preamp pairing can make a big difference in how your microphone sounds.

 You'll find more information about mic preamps in Chapter 3 of this book.

Mic Clips

Most tubular-shaped mics come with a mic clip, which holds the mic and attaches to a stand (Fig. 2.12).

Figure 2.12 The On Stage MY110 mic clip attaches mics to stands.

It never hurts to have a few spare clips on hand because these can break fairly easily. Mic clips are available in quick-release models (Fig. 2.13), which are ideal for vocalists who like to hold the mic in their hands but also trade off with using it on a stand.

Figure 2.13 A quick-release mic clip isn't just for singers who alternate between holding a mic and putting it on a stand, but for when you need to match multiple mics with multiple stands.

Mic Sleeves

These may not be all that useful in the studio, but if you feel the need to accessorize your mic to match your stage clothes with a bit of bling, mic sleeves will do the job (Fig. 2.14). They can even dress up your stand.

Figure 2.14 Give your mic a new suit with mic sleeves, like this one from MicFX.

Stereo Bar

Nothing makes stereo-miking setups easier than a good stereo bar designed to hold and position the two mics. These are available in a variety of sizes and configurations (Fig. 2.15).

Figure 2.15 The DPA Microphones SB0400 holds two mics, but the company also makes a five-mic tree mount for surround.

Typically, two movable components position the microphones as desired, and the bar attaches to any mic stand with a 3/4-inch threaded connector. It's also possible to obtain bars for more than two mics.

Remote Control

If you're engineering yourself, you may be sitting fairly close to your computer and dealing with fan interference and other computer-related noises being picked up by your mic. Or, if you've set up your mic far enough away to avoid noise, you may have to run back and forth between the mic and the interface to tweak levels, as well as returning to your recording software to make any needed settings. Fortunately, several programs offer remote control applications for iOS and/or Android devices (Fig. 2.16).

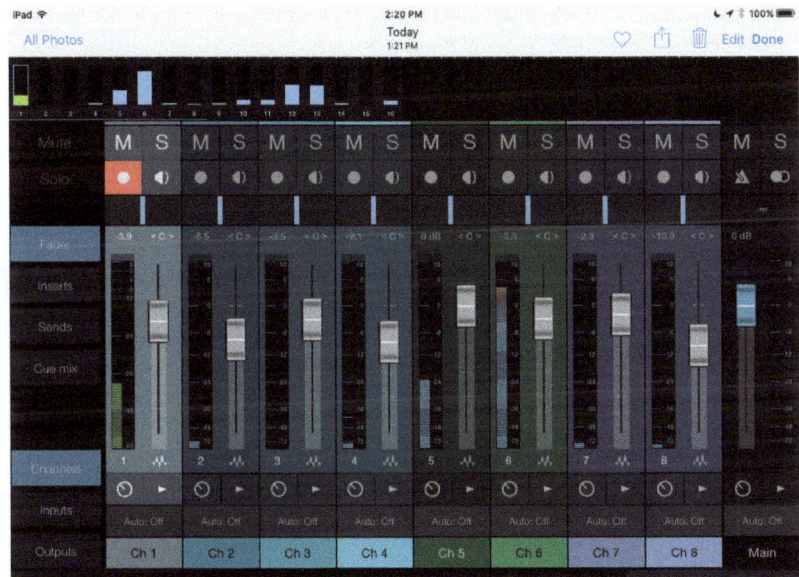

Figure 2.16 This remote application for PreSonus Studio One allows running the transport and setting levels remotely, which simplifies recording away from noise sources.

If your software doesn't have a remote control application, one option is to use a wireless QWERTY computer keyboard and learn the keyboard shortcuts needed for recording (record, stop, etc.). You can typically be at least 20 feet away, even if there's a wall between you and the wireless keyboard receiver (usually a small device that plugs into a computer's USB port).

Key Takeaways

♦ High-pass filters, which are often built into a mic, can be engaged to reduce low-frequency response. This can help reduce the proximity effect and pickup of sounds below the range of the instrument being miked.

♦ A pad button reduces a mic's sensitivity so it can record higher-level sounds without distorting.

♦ Acoustical screens and shields minimize room reflections from entering back into your mic after bouncing off walls. They can also help shield a mic from computer fan and hard drive noise.

♦ Mics are sensitive to physical shock, dust, and moisture. You can protect them with cases, pouches, or even just a plastic bag when not in use.

♦ A pop filter is essential when close-miking singers, where *p* and *b* sounds cause audible pops that are difficult to remove from a recording.

♦ Most professional mic cables have balanced, XLR-style connectors at each end.

♦ Mic stands, often with a mic boom, are essential for recording instruments.

♦ Shock mounts help prevent mics from picking up vibrations transmitted from the floor that travel up the mic stand and into the mic.

♦ Mic preamps are needed to bring most microphone outputs up to a useable level.

♦ A stereo bar holds two microphones in place to simplify some stereo miking techniques.

Chapter 3

Microphone Connections

Now that we've covered mic types, polar patterns, and accessories, let's look at all those numbers that manufacturers throw at you in ads and spec sheets. You don't need to become an expert on electrical details—we'll cover only what you need to know to connect and use mics properly.

Impedance

Impedance is the resistance that an electronic component or device has to the flow of alternating current—think of it as the electrical equivalent of friction. It's measured in ohms and can be represented in several different ways. Impedance can be indicated by the following:

- A numerical measurement specified as so many ohms

- The Greek Ω (omega) symbol

- The letter Z (as in Low-Z or High-Z, short for low impedance or high impedance, respectively)

 Impedance can relate to both inputs and outputs.

Impedance Options

Mics have two basic output impedances: low impedance (50-600 ohms) and high impedance (above 10,000 ohms, or 10 kohms). The two are not interchangeable without impedance-matching transformers; without boring you with the math, the basic idea is to use low- or high-impedance microphones only with inputs designed to accommodate one impedance or the other.

Preamps and mixers designed for recording are almost always designed for mics with low-impedance outputs. Some old guitar amps, which have high-impedance inputs, included a mic input jack for high-impedance mics. For recording, you'll want to choose mics with low-impedance outputs. However, Shure does make a high-impedance harmonica mic, the 520DX, under the assumption that most players will plug it into a guitar amp with a high-impedance input.

Many computer audio interfaces feature a high-impedance input designed for plugging in guitars and basses with passive, low-level pickups. This input is also suitable for high-impedance mics.

Some mics (Fig. 3.1) offer a switchable output impedance, so they can work with low- or high- impedance inputs. If your mic offers this feature, make sure it's set for low impedance unless you expect to use a high-impedance input.

Figure 3.1   The Shure 545 SD has an internal jumper inside the mic that selects either a low- or high-impedance output.

Why Low-Impedance Mic Outs Are Better

For recording and live sound, avoid high-impedance models. Low-impedance mics can drive longer cable runs without the risk of interference (picking up hum and noise). They also help avoid the high-frequency signal loss associated with high-impedance mics and long cable runs. Most high-impedance mics are also lower quality than low-impedance types. Nearly all professional recording mics are low-impedance models.

Although the input impedance of computer audio interfaces, mixing boards, and mic preamps is generally five to ten times higher than the mic's output impedance (usually in the 1,000–2,000 ohm range), this results in optimal performance for the pair. Plugging into a jack with an input impedance equal to or higher than the mic's output impedance avoids signal loss.

Microphone Levels

Different types of audio signals operate at different levels, and unless you use a device to convert them, those levels are not interchangeable and compatible. Mixers, synthesizers, audio interfaces, CD players, and similar amplified electronic devices have *line-level* outputs that are much higher level than *microphone-level* signals. Except in very rare cases, your mic's signal level will not interface properly with line-level equipment until it's amplified by a microphone preamplifier (*mic pre* or just *preamp* for short). You'll find preamps built into mixing boards, audio interfaces, and even as stand-alone units. They all boost the mic level signal to a line level that's suitable for other devices in your system.

Sound Pressure Levels

It's important to know the maximum loudness, or sound intensity, that a mic can record. Sound intensity is referenced to a specification called the Sound Pressure Level, or SPL. SPL is expressed in decibels.

Tech Talk: Decibels

Decibels refer to a ratio of one sound level to another and therefore require a reference level to establish the ratio. With SPL, the reference level is usually based on the threshold of human hearing. Decibels are not a linear scale, but a logarithmic scale; doubling a decibel ratio doesn't represent a doubled value, but a multiplied one. This is necessary to accommodate the huge range of numbers that we'd otherwise have to write out when dealing with the difference between the softest and loudest sound levels the human ear can perceive. For example, a ratio of 10 dB means one signal is about ten times more intense than another signal, while a ratio of 30 dB means it's about 1,000 times more intense.

Mics can handle a wide range of sound pressure levels, and you'll often see a specification for their maximum SPL. This is the loudest sound level they can capture without audible distortion; it usually represents the point where distortion reaches 0.5%. A microphone with a maximum SPL specification of 130 dB can handle much louder signals without distortion than a mic that can handle 120 dB SPL. In actual use, either one would probably serve you well, but it depends on the loudness of the sound sources you want to record.

Most sound sources in a home studio environment will be below 130 dB SPL. However some sounds, such as a very loud guitar amp, a drum kit being played with power and enthusiasm, or a trumpet may hit that level when close-miked. By way of comparison, a jet engine generates sound pressure levels in the 140 dB range, even when measured from fifty meters away. Some other typical SPL figures include 40 dB SPL for background noise level in a library, 50 dB SPL in a quiet residence, 60 dB SPL for average human speech from 1 meter away, 70 dB SPL for a vacuum cleaner at 1 meter, 110 dB SPL for a chain saw at 1 meter, and 110 to 115 dB SPL for an orchestra or loud rock concert.

 High sound pressure levels can damage your hearing. Exposure to noise levels between 85 and 94 dB SPL for extended periods can cause hearing loss. Above 97 dB SPL, even short-term exposure can be dangerous.

A sound level meter (Fig. 3.2) can help you determine whether you need to use hearing protection while recording.

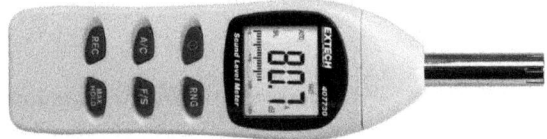

Figure 3.2 The Extech 407730 Digital Sound Level Meter measures sound levels from 40 to 130 dB, as registered on a large LCD display or analog bargraph.

Microphone Sensitivity

Even when placed the same distance from a single sound source, different microphone models will often have different output levels because they have different *sensitivity*. The more sensitive a mic, the better it is at converting small acoustic pressure changes into output voltages. More sensitive mics require less preamplification because they generate higher output levels.

A microphone's sensitivity is measured in millivolts per pascal at 1 kHz. *Wait!* Don't turn the page! All you need to know is that the more negative the number, the less sensitive the mic. For example, the Shure SM57 moving-coil dynamic microphone has a sensitivity spec of –56 dBV/Pa. This is a less sensitive microphone than Shure's SM81 small-diaphragm condenser mic, whose sensitivity spec is –45 dBV/Pa.

Most condenser microphones have a higher output level compared to moving-coil dynamic mics, as the above comparison shows. In general, ribbon microphones have notoriously low sensitivity—even lower than most moving-coil dynamic mics. Their low output levels require microphone preamps with sufficient *gain,* which we cover next.

Preamplifiers

Mic preamps are in all sorts of musical equipment—mixing boards, computer audio interfaces, and dedicated external mic preamp units such as the Black Lion Audio B12A MKII (Fig. 3.3).

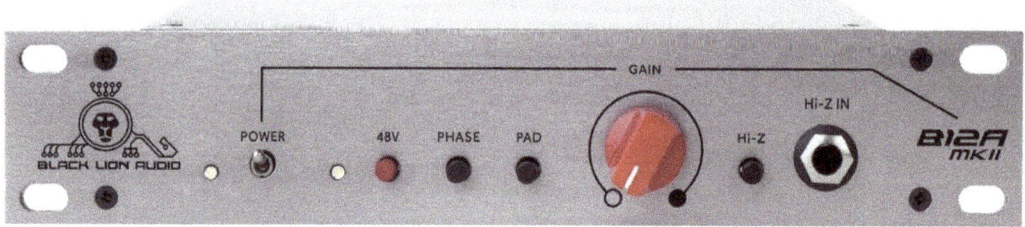

Figure 3.3 The Black Lion Audio B12A MKII microphone preamplifier not only offers an XLR mic input with 48V phantom power on the back, but also a high-impedance input for guitar or bass, phase-reverse button, and pad (level-reducer) button.

Mic preamps are available in single-channel, stereo (two-channel), and even 4- to 8-channel formats. Mic preamps are also an essential element of multi-processor *channel strips,* which usually include a preamp, compressor, and equalizer in one housing.

While the features and available gain of mic preamps vary considerably, all are designed to bring up the low mic-level output signals to line level.

The sound quality you'll hear from your mic depends upon the quality of the preamp it feeds. Fortunately, technology has progressed to a point that even low-cost audio interfaces offer good quality preamps. The marketing departments of different manufacturers will wax poetic about their "clean, warm, transparent" preamps, but the reality is that most modern preamps in the same price category have very little difference, ranging between "excellent" and "somewhat more excellent."

Some mics, notably USB microphones like the Blue Yeti Pro and IK Multimedia iRig Mic Studio (Fig. 3.4) have built-in mic preamps. These receive power from the USB bus and do not need an external preamp.

Figure 3.4 The IK Multimedia iRig Mic Studio USB microphone doesn't need a preamp or XLR cable, because it connects directly to your computer through a USB port.

Let's look at preamp specs and features.

Gain

The amount of available gain is provided by a preamp specified in dB. The higher the number, the greater the potential gain—e.g., a preamp with 60 dB of gain can provide more amplification than a preamp with 50 dB of gain. When close-miking instruments, just about any amount of gain will be enough. You'll need more gain for room mics, whispering, and quiet performances.

For most microphones, 50 to 60 dB of gain is sufficient. Ribbon microphones, which used to be rare but are becoming more common thanks to improvements in modern manufacturing techniques, require as much as 70 dB of gain for quiet signals. Few audio interfaces can provide that amount of gain (check the specs), so you can either turn up the preamp gain as much as possible and hope that the signal-to-noise ratio is good enough, or use an external, high-gain preamp and connect its output to the audio interface's line input.

Preamp Class Types

Preamp circuit design has various classifications. The two most common types are Class A and Class AB. One class isn't inherently better than the other; a well-designed Class AB preamp with quality components will sound better than a poorly designed Class A preamp with cheap components. And of course, how a preamp sounds to *you* is ultimately what matters—your personal preference may lean toward a totally clean and neutral preamp, or toward one with some character.

That said, all things being equal, a Class A preamp will deliver more accurate amplification than a Class AB type, which is designed for efficiency. (Efficiency is why Class AB is commonly used for power amps; Class A amplifiers draw more power.)

Traditionally, some favorite preamps were those with a certain characteristic sound instead of those that were totally neutral. However, as digital technology has become ever more refined and plug-ins reproduce ever more nuanced sonic characteristics, the trend is toward neutral preamps that add no character of their own. That way, any character can be added only by plug-ins and choices made by the engineer—it's analogous to why painters start with a white canvas.

Mic Preamp Specifications

While audio interface specs usually apply to the interface as a whole, the mic preamps are the most crucial factor in determining these specs. *Spoiler alert:* Most manufacturer spec sheets are meaningless, because they don't provide the conditions under which the tests were taken.

For example, the results for testing just about anything will be different depending on whether the preamp gain is set to zero or set to maximum. Results can also depend on the reference level—for a given amount of noise, if you're measuring signal-to-noise ratio by comparing the maximum possible output level to the residual noise, that spec will seem much better than a reading taken with a reference level that's lower than the maximum possible output. With that in mind, let's look at the meaning of five important specs.

Frequency Response

Ideally, audio gear designed for maximum accuracy should reproduce all audible frequencies equally—bass shouldn't be louder than treble, or vice-versa. A frequency response graph measures what happens when you feed test frequencies with the same level into a device's input, then measure the output to check for any variations. You want a response that's flat (even) from 20 Hz to 20 kHz, because that's the audible range for humans with good hearing. It becomes harder to reproduce extremely high or extremely low frequencies. Fig. 3.5 shows a typical frequency response graph.

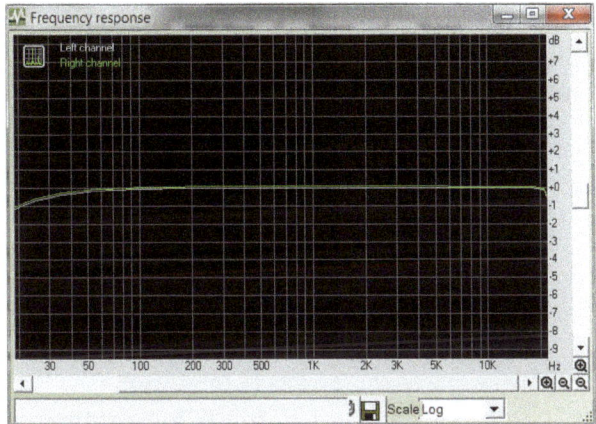

Figure 3.5 This shows a frequency response that has essentially no response deviations from 50 Hz to 20 kHz, and is down 1 dB at 20 Hz.

As with most preamps, the response goes down even further below 20 Hz; this is deliberate, because there's no need to reproduce signals we can't really hear. So, this graph shows that the preamp is quite accurate when reproducing signals at audible frequencies.

A manufacturer's specification will give a frequency range and amount of response deviation. For example, the spec for the above interface could be expressed as –1 dB, +0 dB from 20 Hz to 20 kHz, or ±0.5 dB from 20 Hz to 20 kHz.

Signal-to-Noise Ratio

All electronic circuits generate some noise, so you want the lowest possible noise level. Noise increases as you turn up the gain. For example, here's a graph of a mic preamp's noise with the volume turned up one-fifth of the way (Fig. 3.6).

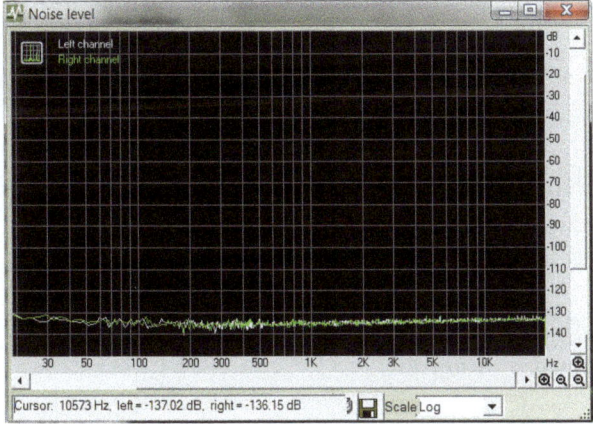

Figure 3.6 The noise is less than –130 dB. In other words, compared to a signal at full level, the noise is over 130 dB softer.

The graph in Figure 3.6 shows that the signal-to-noise ratio (the ratio of the full-level signal to the noise) is 130 dB, which is very quiet. But if we turn up the mic preamp gain two-thirds of the way—about right for recording a quiet vocalist—the noise level increases, approaching –110 dB (Fig. 3.7).

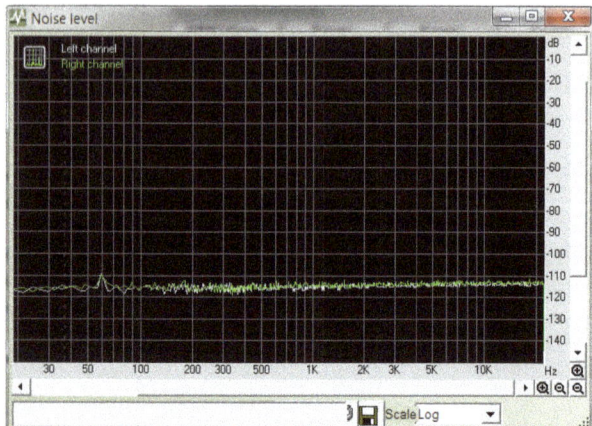

Figure 3.7 With the gain turned up, the noise increases but is still extremely quiet. Consider that Compact Discs can at best reproduce levels no lower than about –90 dB. Essentially, the noise is so low that a CD can't even reproduce it.

These two graphs also hint that specs have the potential to mislead rather than enlighten. *Company A* might use a graph like the upper one for their marketing, while *Company B* might choose a more real-world graph like the lower one. Company A could imply that their mic preamp is quieter—"just look at the specs!" So it's important to know the conditions under which specs are taken, and of course, these usually aren't given (and to be fair, most people wouldn't know how to interpret them anyway).

Noise specs are also presented in different ways, so it's important to compare the same kind of specs. The noise spec may be given as *Equivalent Input Noise* (EIN), which represents the noise a mic preamp adds to a microphone's signal. Lower numbers are better. These numbers are very small, which make the preamps look good. Another option is *signal-to-noise ratio*, which compares the ratio of a specific level signal (usually for a given, low level of distortion) to the residual noise. For example, a signal-to-noise ratio of 90 dB means the signal is 90 dB louder than the noise.

The noise spec may have a particular *weighting* that excludes noise not relevant to our hearing. Another way to specify noise, *dynamic range,* is the ratio of the maximum possible signal to the residual noise.

Total Harmonic Distortion

Just as all circuits generate noise (no matter how small), they also generate distortion. A Total Harmonic Distortion graph shows the level of the harmonics generated by distortion.

This test feeds in a 1 kHz signal at maximum level. In theory, the output should consist only of that 1 kHz signal. Any other signals represent distortion. Noise will often mask this distortion, so turning gain all the way down to minimize noise shows distortion components more clearly (Fig. 3.8).

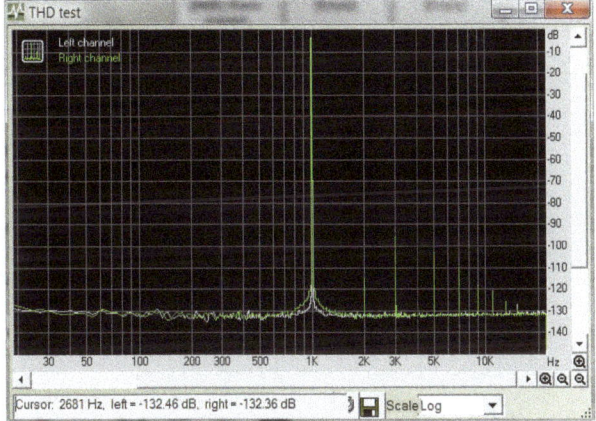

Figure 3.8 With the preamp gain all the way down, the distortion components at 2 kHz and above are clearly visible above the noise, yet still very low.

Intermodulation Distortion

Intermodulation distortion occurs when two signals interact in such a way that they produce artifacts. These artifacts also represent distortion, and many people consider intermodulation distortion more objectionable than harmonic distortion. This test feeds in two signals at maximum level, one at 60 Hz and one at 7 kHz. Any output signals other than these two frequencies represent distortion (Fig. 3.9). This graph of intermodulation distortion has the gain turned all the way down to minimize noise.

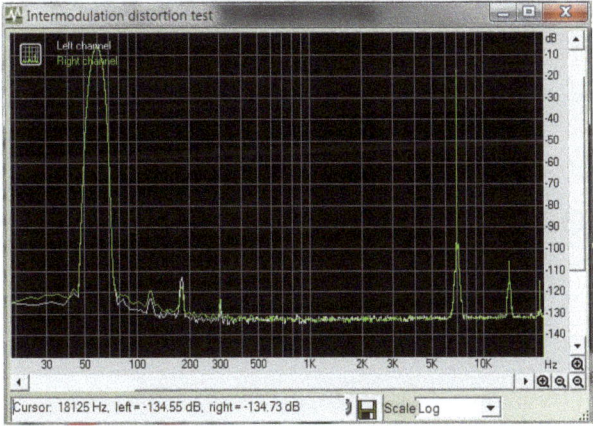

Figure 3.9 With no gain, you can see low-level distortion below −110 dB.

Crosstalk

Crosstalk occurs when one channel picks up a signal from another channel. This happens because some circuit elements radiate signals, while other elements pick up those signals. Careful mechanical design and signal isolation can reduce crosstalk, but you can't eliminate it entirely. Crosstalk (Fig. 3.10) is more likely with high gain, high frequencies, and sometimes also affects low frequencies. Minimizing crosstalk is especially important when you're recording stereo sources.

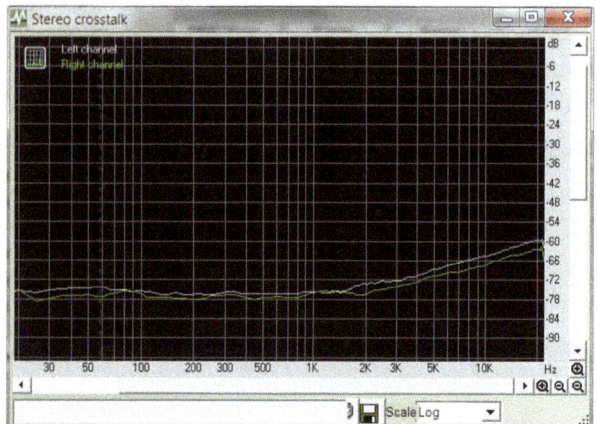

Figure 3.10 This crosstalk spec is taken with a relatively high-gain setting, as you might use when miking an instrument in stereo. With lower amounts of gain, the crosstalk goes way down.

The graphs in this section illustrate an important point: Specs can be a snapshot of a gear's particular state and can be represented by a range of numbers. Always use specs as a guide, not a judge—and understand that the specs you read are often fairly meaningless.

 Rane has published a very deep explanation of specifications, available on their website at http://www.rane.com/note145.html.

Connecting Your Microphones

How you connect your microphone to a mixer, audio interface, or preamp depends on whether the mic is a dynamic, condenser, ribbon, or tube mic.

Moving-Coil Dynamic Mics

With low-impedance, moving-coil dynamic mics, the connection process is simple. You'll need an XLR male to XLR female cable or very rarely, a female XLR to 1/4-inch TRS cable. While not nearly as common as XLR inputs, a few audio interfaces use TRS jacks (Fig. 3.11) for their low-impedance mic inputs.

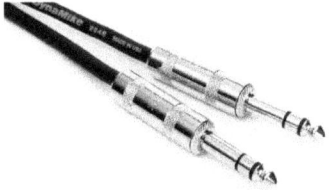

Figure 3.11 TRS connectors appear superficially like standard guitar cable plugs, but look closer—there's an extra *ring* conductor between the *hot* conductor and ground.

Tech Talk: 1/4-Inch TRS Connectors

1/4-inch balanced TRS phone jacks appear like standard 1/4-inch guitar cord-type jacks, but use stereo jacks that are wired for balanced operation (like XLR jacks) instead of being wired for stereo. Because it's not obvious from looking at the outside of phone jacks whether they're balanced or unbalanced, manufacturers often indicate that they're balanced (and may even show the wiring) on the gear's panel that holds the jacks.

Low-impedance mics usually have a male XLR output jack, while mic preamps usually have a female XLR input jack. The female end of the XLR cable connects to the mic, and the other end of the cable connects to the mic input on the mixing board, computer audio interface, or preamp. Most ribbon and condenser microphones connect the same way, although there are some important considerations for powering these mics that we cover later in this chapter.

Condenser Microphones

These connect in the same way as typical dynamic microphones, although for most mics you'll need to enable 48V phantom power to provide power for the mic's electronics (as described later).

Tube Microphone Connections

As alluded to earlier, mics with tube electronics usually come with their own separate power supply unit because tube circuits require more voltage and current than phantom power can provide. Unlike dynamic mics, they typically require two cables:

- ◆ A specialty, multi-pin cable carries power from the AC-powered power supply to the mic's electronics, as well as audio from the mic back to the power supply.

- ◆ A second, standard XLR male to XLR female mic cable routes audio from the power supply to your audio interface, mixing board, or mic preamp.

USB Microphone Connections

USB mics use the same cables as any USB peripheral. However, the mic may not work properly unless patched to a USB port that connects directly to the computer's motherboard. Using a *USB hub* (a device that plugs into a computer's USB port and splits it into multiple USB ports) is not recommended. USB mics will generally work with USB 2.0 ports and are compatible with USB 3.0 ports, or with USB-C ports using an adapter.

Powering Condenser and Ribbon Mics

All condenser mics require a power source. Their output levels are so low that they need an internal preamp that raises the signal to mic level, after which a second, external preamp boosts the level further. Nearly all condenser mics have built-in electronics to provide the needed initial stage of amplification.

Some condenser mics power their internal electronics with a battery or two; these mics will have an onboard battery compartment and (usually) a built-in switch for turning the battery on or off. This solution is great for *field* use (recording remotely where no AC power is available). It's also acceptable for studio use, as long as the mic uses standard battery types and has a long battery life to minimize session interruptions. Before considering any battery-powered mic, make sure the battery is easy to obtain and to replace.

Because dealing with batteries can be a hassle, audio engineers developed *phantom power* to send power to the mic over an XLR cable. This doesn't interfere with the mic's audio output signal. How it works is less important than recognizing that a condenser mic without onboard batteries or some alternative power source won't work without phantom power. If you connect a condenser mic and it doesn't seem to be working, check to see if phantom power is turned on. Your audio interface may have a switch to enable phantom power for individual inputs or for groups of inputs.

Tech Talk: Phantom Power

Phantom power, invented by Neumann microphones, sends 48V through an XLR cable by way of the wires that go to pins 2 and 3, with a resistor between the power source and each cable (Fig. 3.12). Because these same two wires carry audio to a mixer or audio interface's mic preamp, a capacitor blocks this DC voltage from entering the mic preamp electronics.

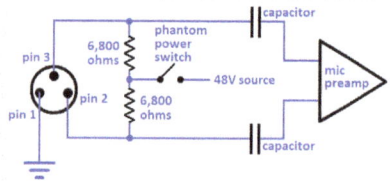

Figure 3.12 This schematic shows a typical phantom power implementation.

Although mics specify 48V for operation, this voltage isn't too critical. Most condenser mics can operate at 35V or even less. However, you can't count on this being the case. At too low a voltage, the mic may not operate at all, or it may have reduced audio performance.

Ribbon mics typically need the most gain of all, but some ribbon microphones (such as the Cloud 44-A and Royer R-122 MKII shown in Fig. 3.13) have a built-in preamplifier that allows using them with standard mic preamps. Otherwise, you normally need about 60 to 70 dB of gain.

Figure 3.13 The Royer R-122 MKII was the first ribbon mic that took advantage of phantom power to provide power to an onboard preamplifier, thus allowing it to connect like a standard condenser mic.

Occasionally you'll hear dire warnings about how phantom power can kill a passive (unpowered) ribbon mic and destroy the sensitive (and expensive) ribbon element. While this is true for some older models, it's not likely with most modern ribbon mics. But to be safe, disable the phantom power on any mixing board, audio interface channels, or external mic preamp before plugging in a ribbon mic. The greatest risk of phantom power damage occurs when connecting or disconnecting a ribbon mic.

 Note that the default setting on some audio interfaces has phantom power turned off. You need to enable it for condenser mics as needed when the audio interface powers up.

After connecting a ribbon mic, it's *usually* safe to turn the phantom power back on (if needed for a group of mics that includes a condenser, for example). When you're ready to disconnect the ribbon mic, be sure to turn off phantom power before unplugging.

Phase and Polarity

Like waves in water, sound waves traveling through the air have a peak (high point, where the air molecules are most heavily compressed) and a trough (low point), where the air molecules are more rarefied. And like waves on a lake, when one sound wave's peak meets another sound wave's trough, their energies cancel each other out to some extent. The more similar the two waves are in size and energy and the more precisely the peak and trough align, the more completely they'll cancel.

When you mic a source with two microphones, unless they are exactly the same distance from a sound source, one might pick up a peak from a wave while the other picks up a trough. If you listen to either mic by itself, it will sound fine. But if you mix the two together and listen in mono, any cancellation will create a composite sound that's weaker and "hollower" than the sound of either mic by itself.

Recall that balanced XLR mic cables have two wires with signals that are out of phase with each other, and this is why the balanced line can reduce noise. Most mixing boards and outboard mic preamps have *phase* switches (also called *polarity* switches) that reverse the two wires, thus reversing the polarity. This can also be done electronically in most recording software. Sometimes products indicate this switch or function with a "slashed zero" symbol – Ø (Fig. 3.14).

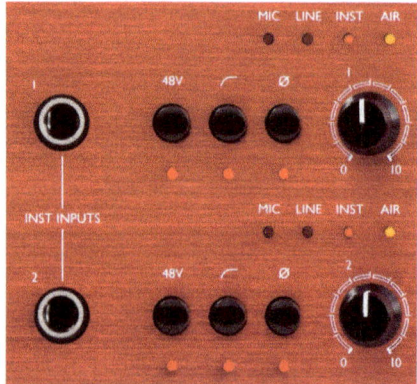

Figure 3.14 The Focusrite Clarett 8PreX has front-panel, hardware switches for a high-pass filter (to the right of the 48V phantom power switch) and phase (to the right of the high-pass switch).

When using multiple microphones in close proximity, you'll need to listen carefully in mono while flipping the polarity button on one of the two channels to decide if this improves the sound. Out-of-phase signals will usually sound weaker and thinner, with noticeably reduced bass.

Also note that some people make sure the polarity is the same as the source—for example, when miking a kick drum, pushing on the kick drum beater pushes air outward from the drum's head. To do this, check that the speaker cone pushes outward when the kick drum hits. If not, consider reversing the phase.

Although many people will tell you there's no sonic difference, others believe there is. Listen for yourself, and then decide based on your own experience.

Key Takeaways

◆ Most microphones are low-impedance types and are designed to work with preamps that accept low-impedance outputs.

◆ To avoid a loss in output level, the preamp input impedance should be about five to ten times higher than the mic's output impedance.

◆ Microphones have limits with regard to the loudness or sound pressure level (SPL) they can record. Make sure the mic you choose can handle the sound level you want to record.

◆ Microphones also have different sensitivities. More sensitive mics are needed for low-level sources like nylon-string guitars.

◆ Mic preamps need to deliver at least 50 dB or so of gain. 60 dB allows for recording quieter sounds more easily. Traditional ribbon mics require 70 dB of gain or even more.

◆ A well-designed Class A preamp will have lower distortion than a well-designed Class AB preamp, but the Class A preamp will require a power supply with more current.

◆ Preamps are rated by specifications like noise, distortion, and crosstalk. However, these specs are often meaningless, because manufacturers use different testing protocols that may or may not be made clear from the spec sheets.

◆ Condenser mics require a power source, typically supplied as phantom power from a mixer, audio interface, or stand-alone mic preamp.

◆ Tube microphones need a separate power supply due to requiring a higher voltage than phantom power can provide.

◆ Many preamps have switches to reverse the audio's phase (polarity). When using multiple mics, changing the switch position can reduce phase cancellations that could result in a weaker, quieter sound.

Chapter 4

The Home / Project Studio Microphone Cabinet

Outfitting a home or project studio with a balanced, useful assortment of microphones is daunting. With so many types and models available, it's difficult to decide which ones will serve you best. Collecting microphones is a lifetime endeavor for most recording engineers, and you'll likely augment your collection over the years. In this section we'll look at some considerations, and make some recommendations for a basic microphone collection that can capture a variety of sound sources.

Condenser Mics

Large-diaphragm condensers tend to be the mic locker's stars. They work on practically anything, although they're often used for a recording's featured elements: drum overheads and room mics, piano, acoustic guitars, lead and backing vocals, strings, brass, reeds, and guitar amps. They are also useful for stereo miking applications (Blumlein, mid-side, and spaced pairs). However, sound arriving off-axis can sound colored, and some large-diaphragm condensers can sound too "hyped" in the high frequencies.

 Stereo miking techniques are discussed in Chapter 5 of this book.

Small-diaphragm condensers are equally versatile. They're a popular choice for drum overhead mics and general stereo-miking applications—especially for spaced and XY coincident pairs. Other uses include acoustic piano, acoustic guitars, hi-hats, small percussion items, and stereo ensemble recordings (especially choirs). They generally sound more realistic than large-diaphragm condensers, due to the more natural off-axis response. However, they do tend to have more self-noise than their larger condenser cousins. Some models will overload when used up close on very loud sound sources. Pad switches that lower their sensitivity can be useful in those cases.

Tech Talk: How Condenser Mics Work

Condenser is another word for "capacitor," a common electronic component. Here's how the condenser mic works (Fig. 4.1).

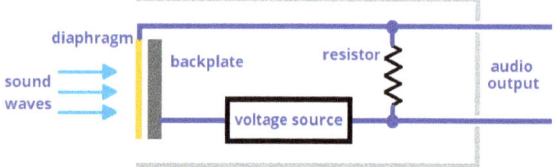

Figure 4.1 A condenser mic creates a voltage from capacitance changes.

The diaphragm is a thin, low-mass membrane that can respond to minute air pressure changes. The diaphragm needs to be electrically conductive, and the usual choice is gold-sputtered mylar. There's a tiny airspace between the diaphragm and an electrically charged backplate (which can be powered by a battery or 48V phantom power). These two components form a capacitor. When the diaphragm moves, the amount of capacitance changes. This generates an electrical signal.

Although the capacitive changes produce a reasonable amount of voltage, the capacitance is so small that it generates very little current. The onboard electronics buffer the signal through a very high resistance, and amplifies it sufficiently for a subsequent mic preamp.

The reason a condenser mic is more accommodating of high frequencies than a dynamic mic is because the diaphragm is much lighter than the diaphragm used in dynamic mics.

Ribbon Mics

Ribbon mics are somewhat more specialized. They excel as guitar amp mics, drum overheads, and room mics. They're also good on brass and sax, as mics for a Leslie speaker cabinet's horn speakers, and even for instruments like banjo and fiddle. On vocals, the tonality can be smooth, detailed, and with a somewhat old-fashioned or retro vibe. However, you do need to watch out for windblasts from instruments and singers—these can damage the ribbon element, particularly with older models. Ribbon mics also exhibit a heavy proximity effect when close-miking.

Tech Talk: How Ribbon Mics Work

As with other mics, a ribbon mic requires an element that responds to variations in air pressure, and a way to convert those variations into voltages that reflect the variations (Fig. 4.2).

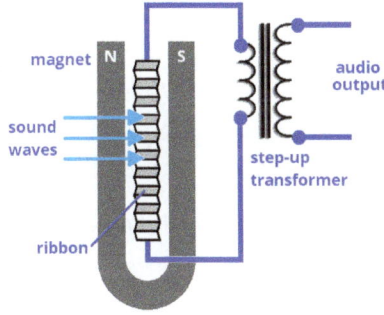

Figure 4.2 A ribbon mic is potentially fragile, but offers an inherent figure-8 response.

This is a variation on the dynamic mic, because it suspends an incredibly thin corrugated metal strip (typically aluminum, as fine as 1/50th the thickness of a human hair) between two poles of a magnetic structure. Unlike dynamic mics that respond to air pressure, the ribbon mic element responds to air particle velocity. As the metal strip moves within the magnetic field, it serves as both the mic's diaphragm and its transducer, and generates a tiny voltage. A transformer increases the level and converts the impedance to make it suitable for conventional mic preamps. Ribbon mics provide a figure-8 response because they respond to air velocity variations hitting the ribbon's broad front and back, but not its thin sides.

Dynamic Mics

Large-diaphragm dynamic mics are a good choice for recording loud, low-frequency sound sources—kick drums, floor toms, bass amps, low-frequency Leslie speakers, and lower-register brass instruments like trombone and flugelhorn. They're also popular for broadcast vocal and narration recording, and with the right singer, can work well for hard-rock vocals. While large-diaphragm models may be more expensive than other moving-coil dynamics, their versatility and low-frequency capabilities warrant having at least one or two in your collection.

Tech Talk: How Dynamic Mics Work

Dynamic mics are known for being rugged and also quite reliable. However, it's possible to make sensitive dynamic mics; here's how they work (Fig. 4.3).

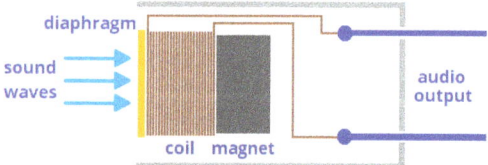

Figure 4.3 The dynamic mic is a favorite for live performance, but it also has a place in both traditional and modern studios.

A dynamic mic's diaphragm attaches to a wire coil that can move back and forth on a magnet. The more air pressure strikes the diaphragm, the more the coil cuts across the magnetic field, which generates a voltage in the coil. This voltage becomes the audio output.

Because the diaphragm is less fragile than the kind found in a condenser or ribbon microphone, it can handle higher-level signals. The tradeoff is lower sensitivity and slower transient response.

Some models, designed specifically for kick, may have a *scooped* (reduced gain) lower midrange response, which could make them less versatile overall. Others, such as the RE320 (Fig. 4.4), offer switch-selectable responses for both kick and general-purpose use.

Figure 4.4 The versatile Electro-Voice RE320 large-diaphragm dynamic mic has a switch that can optimize the frequency response for miking kick drums.

Small-diaphragm dynamics are the modern mic locker's workhorses. They work well on snare, rack toms, congas, bongos, djembe, and other smaller drums, and are the go-to guitar amp mic for many engineers. For placements where a microphone might get banged or subjected to abuse (like being hit by a drum stick while recording), small-diaphragm dynamics usually get the call due to their ruggedness and low cost. The downside is that compared to condensers, their response generally doesn't reach as far into the highest and lowest ends of the frequency spectrum.

Outfitting the Mic Locker and Building It Up over Time

A well-chosen "baker's dozen" mic assortment will cover 95% of the recording tasks a small project studio may encounter. You'll have enough resources to track a typical, small-ensemble rhythm section of drums, bass, and guitar simultaneously.

But even with a careful selection of versatile mics that hit the budget/high-performance sweet spot, the cost of a dozen microphones is still significant. If you can't afford to purchase everything at once, prioritize and get the basics, then add to it when you can. For example, with drum miking you can start with a basic four-mic drum setup: large-diaphragm dynamic kick mic, small-diaphragm dynamic for snare, and a pair of ribbon mics or small-diaphragm condensers for the overheads.

Even if your audio interface has enough mic preamps and input channels, with four mics you won't be able to mic other instruments simultaneously with the drums. Yet each of those "drum microphones" has applications outside of drum-kit miking. When you're ready to overdub rhythm and lead electric guitar, or record acoustic guitar parts, you can re-purpose the mics for those tasks. For example, a good multi-purpose, large-diaphragm dynamic mic for the kick can do double-duty on bass amp cabinets. Ribbon microphones or small-diaphragm condensers used for drum overheads can be used for acoustic guitar tracking.

When prioritizing, purchase mics based on the recording projects you do most often—then expand from there.

- If you record mainly finger-style acoustic guitar, a pair of small-diaphragm condensers would be a wise place to start.

- If you record mostly voice-over artists, buy the highest-quality, large-diaphragm condenser and large-diaphragm dynamic mics you can afford.

- If you're a guitarist who works alone and rarely records drums, concentrate on a ribbon mic, a small-diaphragm dynamic, and a large-diaphragm condenser for amp-miking duties; add a pair of small-diaphragm condensers for the occasional acoustic guitar overdub.

For a general-purpose studio on a very tight budget, I'd recommend starting with the four-mic drum setup (with the small-diaphragm condenser option for overheads) and a large-diaphragm condenser. This provides the basic tools needed to record practically any instrument by itself. When your budget allows, add a pair of ribbon mics and additional dynamic mics so you can record multiple players simultaneously.

Suggested Models

Ask ten engineers to list their favorite microphones, and you'll see ten different lists. Mic preferences are personal, and getting to know the various models and their sonic characteristics can be a life-long pursuit. Additionally, some of the best models are expensive—especially for top-of-the-line condenser and ribbon mics.

Still, there are bargains to be had. Here are some suggestions for affordable models that perform well, along with a sampling of more expensive models for additional choices.

Affordable Large- and Medium-Diaphragm Condensers

Some affordable microphones in this category include the following:

- Blue Bluebird
- AKG C214
- MXL 4000
- RØDE NT1
- Neat Microphones King Bee
- Neat Microphones Worker Bee
- Sterling Audio ST77

Higher-End Large and Medium-Diaphragm Condensers

Some more expensive microphones in this category include the following:

- Audio-Technica AT4050
- Mojave Audio MA-201
- Mojave Audio MA-200
- AKG C414 XLS
- Neumann TLM102
- Shure KSM44A

Affordable Small-Diaphragm Condensers

For small-diaphragm condensers, the following mics provide good affordable choices:

- Audio-Technica Pro 37
- Shure PG81

- Audix F15

- TASCAM TM-PC1

- MXL 603S

- RØDE M5

Higher-End Small-Diaphragm Condensers

Some of the more expensive small-diaphragm condensers include the following:

- Blue Hummingbird

- Audio-Technica AT4041

Affordable Ribbons

Good quality affordable ribbon mics include the following:

- Cascade Fat Head II

- MXL R144

Higher-End Ribbons

On the higher end of the spectrum, you'll find the following ribbon mics:

- Beyerdynamic M 160

- Royer R-101

Large-Diaphragm Dynamics

For large-diaphragm dynamic mics, the following are all good choices:

- Electro-Voice RE320

- Audio-Technica ATM250

- Heil PR 40

- Shure SM7B

Small-Diaphragm Dynamics

The following are examples of good quality small-diaphragm dynamic mics:

- Audix i5

- Shure SM57

- ◆ Audio-Technica ATM650

- ◆ Sennheiser e 609

- ◆ Heil PR 22 UT

- ◆ Granelli Audio G5790

Small-Diaphragm Dynamic Tom Mics

Some good small diaphragm dynamic mics for toms include the following:

- ◆ Heil PR 28

- ◆ Audix D2

- ◆ Sennheiser e 604

The Illustrated Collection

Let's take a look at a representative mic locker that would do any small studio proud. The total cost for all of these mics at the time of this writing would be a bit over USD $3,000—admittedly, not insignificant. However, each of these would be a useful addition to *any* mic locker, no matter how extensive. These aren't cheap mics you'll eventually outgrow; with proper care, good microphones like these can last a lifetime. With this collection (Fig. 4.5) or something similar, you have the foundation of a microphone collection that can handle a wide variety of tasks and serve you well for decades to come.

Figure 4.5 A baker's dozen selection of various mics can handle the vast majority of project studio sessions.

The thirteen microphones in the Fig. 4.5 group shot are:

- **Top row:** A pair of Audio-Technica AT4041 small-diaphragm condensers flanking a Mojave Audio MA-300 multi-pattern, large-diaphragm tube condenser mic;

- **Middle row:** A pair of Cascade Fat Head II ribbon mics;

- **Lower row, left to right:** a Shure SM57 and an Audix i5 small-diaphragm dynamic mics, an Audio-Technica ATM250 and an Electro-Voice RE320 large-diaphragm dynamic mics, and three Audix D2 small-diaphragm dynamic/tom mics to round out the collection.

Comparing Mics and Preamps

With the relatively low cost of modern recording tools, even modest home studios can afford a variety of equipment. In addition to multiple mics, it's not unusual to see multiple mic preamps—some built into the computer audio interface, a few more built into a small mixing console, and maybe even a few outboard, stand-alone preamps.

This variety of gear provides multiple sonic options, especially when combining different mics and preamps. Unfortunately, evaluating these options can be complicated. You often won't have enough gear to do direct, side-by-side, quick comparisons of the different possible combinations. When you need to plug, unplug, or do anything that takes time to configure, you can't hear a direct, instant comparison. Yet being able to do so, in the context of the music, is an extremely valuable tool for making more informed decisions regarding what sounds and works best for a specific situation (e.g., which mic works best for a particular singer, or which mic/preamp combination is ideal for miking a guitar amp).

Practical Comparisons

Doing proper A/B comparisons requires a little ingenuity, some careful patching, and the occasional willingness to compromise. Depending upon the specific gear you have available, you may need to devise creative routings to compare multiple combinations quickly.

With a multi-channel mic preamp (e.g., the API 3124+), an audio interface with multiple mic preamps, or a mixing console with multiple mic input channels, setting up comparisons among multiple microphones is relatively easy if enough preamps are available.

1. Plug each mic into one of the multiple, identical mic preamps.

2. Set the mics up in the desired location relative to the sound source (and with their diaphragms as close as possible to each other).

3. Run the direct outputs from the multi-channel preamp or mixing console into your audio interface's line inputs.

4. Assign each mic to a separate track in your recording software.

5.   Match the levels as closely as possible (some mics will be more sensitive than others) so the "louder is better" bias doesn't influence your decisions.

If your recording software has an exclusive solo function, soloing one channel automatically mutes the other channels. This makes fast, direct comparisons easy. If exclusive solo isn't available, set up mute groups so that unmuting one automatically mutes the others (Fig. 4.6). Another option is to use a hardware control surface or hardware mixing console to mute and unmute individual channels quickly.

Figure 4.6 You can assign multiple mics/mic preamp inputs to separate tracks in your recording software if your interface has sufficient inputs.

Multiple mic preamp types are more difficult to audition because it's harder to feed one mic to several different mic preamps simultaneously, which is the optimum way to audition the sonic contributions of different mic preamps. You can use mic splitters to split the signal and run it to two or three different preamps, and even simple do-it-yourself parallel splitters can sometimes work. Ideally, though, you'll want high-quality splitters equipped with high-end isolation transformers for best results and minimal audible side effects when splitting the signal and running it to multiple preamps simultaneously. Conventional splitters (such as the Radial Engineering JS2 and JS3) are good choices, but there are even better options.

Two Terrific Tools

If you don't have enough available inputs on your recording interface/mixer/mic preamp to plug in multiple mics for simultaneous comparison, you can use the Radial Engineering Gold Digger (Fig. 4.7). This is a four-channel, passive mic/line switcher that you can use to switch quickly among four mics or line-input sources. All four inputs feed a common output, but it's not a mixer. It's a switcher, so only one input is active at a time. Trim controls on each channel allow level-matching for fairer comparisons, while 48V phantom power switches on each channel accommodate condenser mics.

Figure 4.7 The Radial Gold Digger can switch quickly among up to four mics for instant comparison.

For mic preamp comparisons, the Radial Cherry Picker (Fig. 4.8) can feed a single mic to four preamps simultaneously, then switch among them quickly to make direct comparisons. The input offers phantom power.

Figure 4.8 The Radial Cherry Picker provides instant comparisons among four mic preamps.

Conceptually, the Cherry Picker is like the Gold Digger in reverse. Both units use relays for the switching, have passive signal paths, and don't add noise or degrade the signal in any way.

Combining the two units gives even more flexibility for auditioning what are arguably the two most important equipment elements of the recording chain/signal path. Some mics sound better on some sound sources when used in partnership with specific preamps, and the best way to decide what works best is to hear the choices in context.

By patching the Gold Digger output into the Cherry Picker input, the Gold Digger can switch between up to four microphones (it also can audition multiple line-input sources and different direct boxes), while the Cherry Picker can select from up to four different mic preamps. You can click a button on the Gold Digger to select the mic you want, and then click another button on the Cherry Picker to pick the desired mic preamp. You'll find that testing mics and preamps in various combinations, and instantly hearing the results side-by-side, is superior to other comparison options.

Key Takeaways

♦ Condenser mics tend to be the stars of the typical mic collection.

♦ A mix of small- and large-diaphragm condenser mics can cover an extremely wide variety of recording applications.

♦ Ribbon mics are more specialized, but can excel in situations where a fast transient response and a warm, natural sound are most important.

♦ Ribbon mics can be damaged by windblasts and exhibit the proximity effect when used for close-miking a sound.

♦ Large-diaphragm dynamic mics are a good choice for recording loud, low-frequency sounds (e.g., bass amps, kick drum, and floor toms).

♦ Small-diaphragm dynamic mics are popular for drums because they'll likely survive being hit with a drumstick. They're also popular for miking guitar amps.

♦ A well-balanced mic collection could consist of two small-diaphragm condensers, a large-diaphragm tube condenser mic, two ribbon mics, two large-diaphragm dynamic mics, and several small-diaphragm dynamic mics.

♦ It is beneficial to compare your mics and preamps in order to decide which options are best for specific applications.

Chapter 5

Stereo Miking Techniques

Comparatively speaking, recording in mono is simple: point the mic at your sound source, move it around for the best possible sound, adjust levels, and hit record. Recording in stereo is entirely different, and there are many ways to set up two mics to capture a sound.

A-B (or Spaced-Pair) Stereo

A-B stereo, one of the most common stereo mic techniques, is also known by other names including *spaced-pair* or *time-difference* stereo. There are two different, mostly genre-dependent approaches to A-B stereo, covered later, but we'll start with the classic A-B stereo approach.

Traditional Setup

A-B stereo uses two mics of the same type—typically matched pairs of omnidirectional condenser mics (although some engineers substitute cardioid models to minimize recording ambient sounds). A-B stereo is far more common with multitrack productions than live recordings. The two mics mount parallel to each other, anywhere from several centimeters to several feet apart, and aim directly toward the center of the stage or ensemble to be recorded (Fig. 5.1).

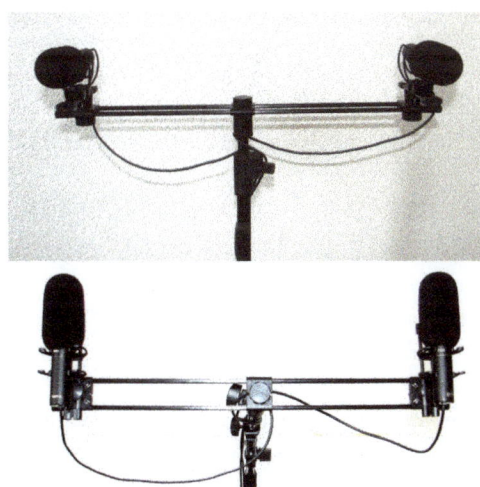

Figure 5.1 The top view shows the spaced pair, as viewed from the sound-source perspective. The bottom view shows what the mics look like when viewed from above.

The distance from the sound source to the microphones is, as with most stereo mic techniques, a matter of taste. Omnidirectional microphones capture more ambience and room sound than directional microphones. Placing them farther back captures more reverb. The optimal placement will depend on the sound source, the room, and your preference for the ratio of direct to ambient sound. The general rule of thumb is to move the mics closer for single instruments and small ensembles in a smaller room, and farther away when capturing a large ensemble in a performance hall. You'll need to experiment to achieve the best balance of stereo imaging, room ambience, and direct sound.

Setup for Classical Music

With classical music, the distance between two mics is generally set between 40cm and 60cm (about 16 to 24 inches). Omnidirectional mics are preferred because their frequency response remains flat, regardless of the distance from the mic to the sound source. Also, unlike cardioid and other directional mics, omnidirectional models don't exhibit a proximity effect, so their bass response stays the same at any distance from the sound source. This allows you to adjust the source-to-mic distance without having tonal and frequency balance shifts.

 Increasing the distance from the source can attenuate high frequencies, although this isn't much of an issue with most real-world mic placement distances.

Directional mics have potential issues with low-frequency accuracy, so they're normally not used when the A-B pair will be the primary source for the entire recording. This is usually the case with classical recordings or live recordings of small ensembles.

Setup for Pop/Rock Music

In recording styles for pop (including rock, hip-hop, alternative, etc.), spaced pairs of directional microphones are much more common, and sometimes with wider mic spacing than typical classical A-B stereo pairs. For example, the Glyn Johns approach to drum miking is essentially a spaced pair of cardioid mics augmented by a kick-drum mic, and occasionally also a snare mic.

 Drum miking and the Glyn Johns approach are discussed in detail in Chapter 6 of this book.

In pop music production, the subdued low-frequency response can be beneficial, because it reduces low-frequency buildup in the mix, so other bass-heavy elements in the mix can compensate. Again, using the Glyn Johns drum-miking technique as an example, the kick-drum mic supplements the kit's primary low-frequency element, and makes up for the attenuated low-frequency response of the distantly placed cardioid mic.

Time Difference Stereo

A-B stereo is sometimes called time difference stereo because this miking technique relies on the spacing between the two mics to generate the stereo image. Our ears are sensitive to small differences in a sound's arrival time at each ear, and these timing differences help our brain determine the directionality of sounds. The spacing of an A-B pair captures these arrival time differences, as well as phase and amplitude differences. The sound from a source will arrive at the mic closest to that source first, then at the second mic a few milliseconds later. The greater the spacing between the mics, the bigger the time difference in signals arriving at each microphone.

Said another way, the wider the spacing between the microphones, the wider the recording's stereo field and imaging. A-B stereo can result in very natural recordings that cover a complete 180 degree arc in front of the microphones, which presents a wide stereo image with a lot of depth. If the stereo soundfield width is of primary importance, then consider A-B stereo a priority.

Disadvantages of Time Difference Stereo

As cool as the resulting stereo image can be, A-B stereo has some weaknesses.

♦ Mono compatibility can be problematic, especially with wider mic spacings and sound sources located in between the two mics.

♦ If the mics are too far apart and too far away from the sound source, the phantom center stereo image can disappear.

♦ Widely spaced mics placed close to the sound source can result in an unnaturally wide and large-sounding instrument in the mix.

♦ Placing the mics too close to one another may compromise the stereo image, although spacing as little as 20 cm can be sufficient if the mics are relatively close to the sound source.

Because of the possibility of phase cancellation when summed to mono, it's important to check mono compatibility by hitting the mono button on your interface or mixing console. If the sound becomes weak and thin when summed to mono, consider compensating for phase issues by adjusting your mic-to-source spacing or the spacing between the two mics. A-B stereo is best for situations where mono compatibility is not a significant concern. With omni microphones, it's also an excellent choice when you need greater mic-to-sound-source distances, such as when recording large ensembles.

XY Stereo

Human hearing is wonderful. Our brain, in conjunction with the spacing between our ears (about 17 to 20 cm), has an amazing ability to localize sounds and pinpoint the direction from which they originate. This is possible due to arrival time differences, where sound arrives at one ear a tiny fraction of a second before it arrives at the other ear, and amplitude (sound pressure level) differences between the two ears.

Amplitude differences are due in part to the baffle effect of the human head, which slightly attenuates the level and changes the frequency spectrum of the sound arriving at the ear is farthest away from the sound source. Our brain's ability to perceive very small differences in sound pressure levels, as well as phase, also gives us the ability to tell which ear is closer and getting the hotter signal. Even very small differences between the times a sound arrives at each ear provide localization information the brain can decipher.

Stereo mic techniques use arrival time differences and/or sound pressure level differences between the microphones to achieve the stereo effect. For example, the stereo image from A-B stereo or a spaced pair of omni mics relies primarily on arrival time differences. By contrast, the XY stereo technique is intensity-based, and depends on the level differences alone.

Required Gear and Setup

The XY stereo mic technique requires a pair of closely matched cardioid condenser microphones, or a single, suitable XY stereo microphone. You will also need two identical microphone preamps with identical settings. XY is a *coincident* mic technique, where the two microphone capsules are as close together as possible, with their center points aligned vertically, but with the microphones angled at 90 degrees relative to each other (Fig. 5.2).

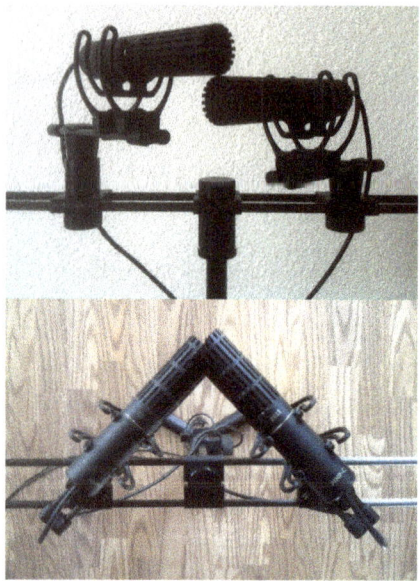

Figure 5.2 The top view shows a pair of DPA 2011C small-diaphragm cardioid condensers configured as an XY pair, as viewed from the sound source perspective. The bottom view shows the XY mics viewed from above.

As shown in the above images, a stereo bar facilitates placing the two microphones. It's possible to use two separate stands and position the microphones correctly, but re-positioning them requires moving both stands. It's much easier to adjust the positioning when both mics are mounted to a stereo bar, attached to a single mic stand.

Tech Talk: Interference Tubes and Condenser Mics

The DPA 2011C microphones seen in some of the figures here are small-diaphragm condenser mics, but they use *interference tubes.* These set the mic capsules farther back down the length of the tube, under the rear-most side grilles, rather than being within a few millimeters of the front (as with most small-diaphragm condenser microphones). When using these particular mics, the ideal configuration would require "crossing" them farther down the tubes so that the diaphragms are aligned vertically. However, this is contrary to the typical arrangement for the vast majority of small-diaphragm condenser microphones (Neumann KM184, RØDE NT5, MXL 604, Oktava MC-012, Audio Technica AT4051, AKG C451 B, etc.). For the sake of clarity, the illustrations show the typical configuration when using mics without interference tubes.

Theory and Cautions

Much of the sound that XY stereo mics pick up arrives off-axis to either one or both mics, so it's best to choose cardioid mics with minimal off-axis coloration. Additionally, directional mics put the proximity effect in play. Placing mics closer to the sound source emphasizes low frequencies more than if the mics are at a greater distance. When setting up XY mics, experiment with the distance from the sound source to achieve the best balance of low frequencies and room ambience.

Because the microphone capsules are essentially in the same physical location, there's no significant difference in the arrival time of the sounds reaching them, regardless of whether they're from the left, center, or right. Therefore, XY miking doesn't provide the arrival-time cues that our ears use to determine a sound's directionality.

However, the capsule arrangement and the attenuation of the cardioid pattern *does* provide some directional cues. The sound will seem to come from the direction of whichever capsule is most on-axis with the sound source (often the louder signal), while the same sound will be more into the null area of the other capsule, thus attenuating it more. So with a proper XY stereo pair of mics, or a purpose-built XY stereo mic such as the one shown in Fig. 5.3, level differences (and to a lesser extent tonal differences) provide directional cues. (This also explains why the stereo effect largely disappears if you use two omni mics instead of cardioid models—there would be neither arrival time differences *nor* any level differences.)

Figure 5.3 The MXL Revelation is a single-unit, stereo XY large-diaphragm tube microphone.

It's possible to use other directional polar patterns in an XY configuration, such as two hypercardioid mics. In fact, the only significant difference between the Blumlein Stereo Pair configuration (described later in this chapter) and XY stereo setups is that Blumlein uses bi-directional mics instead of cardioid models.

XY Stereo Compared to Other Stereo Options

In the section on the Blumlein technique, we'll take a closer look at crossed, bi-directional (or *figure-8*) mics. While both XY and Blumlein stereo techniques use *coincident pair* mics (i.e., mics with diaphragms aligned close to one another), due to the greater attenuation in the bi-directional pattern's null points, the stereo imaging with the Blumlein technique tends to be wider than with XY stereo. The tradeoff is greater ambient- and room-sound pickup.

Using coincident mics also provides excellent mono compatibility for both XY and Blumlein stereo techniques. Mono compatibility is exceeded only by mid-side stereo, which cancels out the side microphone, leaving only the signal from the mid microphone when summed to mono. If mono compatibility is critical, mid-side mic technique is often the preferred choice, but it picks up more ambient sound than XY stereo. If room acoustics are less than ideal, or you want to capture less ambient sound and more direct sound from the source while still maintaining a stereo image, XY is an excellent choice.

ORTF Stereo

ORTF stands for Office de Radiodiffusion-Television Française, a French agency that provided public radio and television between 1964 and 1974. The ORTF stereo mic technique was developed by Radio France in 1960. Unlike other popular stereo techniques such as XY stereo and Blumlein Stereo Pair, which use coincident mics, ORTF doesn't position the two mics as closely together as possible. Instead, they're spaced apart from each other in a *near-coincident* arrangement. Near-coincident techniques usually space the mics at a distance less than that of true spaced-pair arrangements (such as A-B stereo pairs).

Required Gear and Setup

To use this technique, you'll need two identical cardioid microphones. Additionally, a stereo bar and a protractor to measure the angles will help, as will mics with an uncolored off-axis response. For these reasons, small-diaphragm condenser models such as the two DPA 2011C microphones shown in Fig. 5.2 are often preferred for ORTF stereo configurations.

 While it's best to use a matched pair of mics, you can still experiment with the technique as long as the two mics are of the same make and model. Just be aware of any level differences between the two mics, and try to compensate with small gain adjustments at the mic preamps.

To set up for ORTF recording, mount the two mics onto a stereo bar, and angle them outward and away from each other. Set the centers of the two capsules about 17 cm apart from each other, and set the bodies at a 110-degree angle (Fig. 5.4).

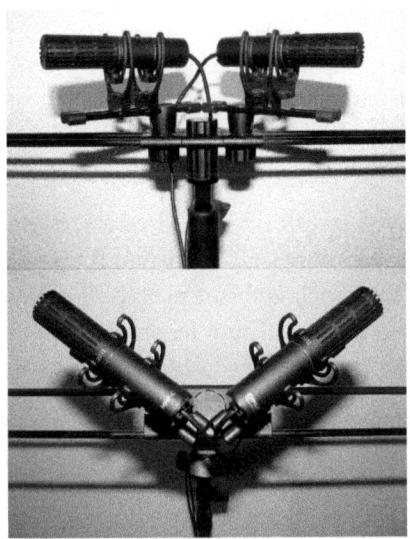

Figure 5.4 The top view shows an ORTF stereo microphone setup from the sound-source perspective. The bottom shows the ORTF stereo pair from above. Note the 110-degree angle of the mics and the 17 cm spacing between the capsules.

This spacing contributes to ORTF's increased realism compared to XY stereo pairs. You'll need to experiment with the distance from the sound source. ORTF tends to pick up less ambience than some of the other stereo mic techniques, so you can often place the mics farther from the source without losing control of the reverberant/direct sound ratio.

ORTF Advantages

Unlike ORTF, XY places the microphone capsules as close together as possible, so sound arriving from any direction arrives at both capsules simultaneously. With ORTF, however, the spacing between mic capsules allows sounds coming from the sides to arrive at one mic a fraction of a second before they reach the other mic. These arrival-time differences provide directional cues for our ears.

While both techniques typically use a pair of cardioid condenser microphones, ORTF stereo gives a wider stereo image than XY. This is because ORTF is a *mixed stereophony* technique that relies on both arrival time and sound pressure level differences from the two mics. The pressure differences result from the angle of the microphones and their cardioid directional patterns. The most sensitive part of one microphone's polar pattern aims to the left, while the second microphone is angled to pick up sounds coming from the right.

When placed in an ORTF or XY configuration, the position of the least sensitive areas of the microphone pickup patterns are contrary to the other microphone. In other words, the most sensitive pickup of one mic covers one side of the stereo image, while the other mic picks up less signal from that side. The two mics reverse roles for the opposite side of the stereo image. ORTF has good mono compatibility, although not as good as true coincident stereo microphone techniques such as XY and Blumlein Stereo Pair.

 Because of the spacing between the mics, you can insert a baffle (e.g., a Jecklin or Schneider Disk) between the mics to further isolate them and increase stereo separation by mimicking the acoustical shadowing effect of the human head. This isn't required or even typical for ORTF stereo setups, but it's another useful option.

Try ORTF setups for smaller ensembles and on individual instruments. Use ORTF instead of XY stereo when you want a wider, more dramatic stereo image, or when mono compatibility isn't as crucial. It's also a good alternative to a Blumlein Stereo Pair when that configuration has too much ambient pickup. If mono compatibility is crucial, do a test recording first to be sure the ORTF recording will sum to mono without objectionable phase cancellation issues. If you find issues, either reposition the mics or consider switching to a coincident pair configuration such as mid-side or XY.

Mid-Side Stereo

Stereo mic techniques are great for capturing and creating a natural or even hyper-natural sense of width and space. Although the two most common techniques are the XY coincident and A-B or spaced-pair configurations, the lesser-known mid-side (M-S) mic technique has unique advantages.

Mid-Side Theory

Some people are hesitant to try mid-side (M-S) recording, maybe because they've read about decoders and math formulas being involved. Don't worry—we'll keep it simple.

This technique requires two mics: one cardioid and one bi-directional (figure-8 response). Ideally, you want similar mics regarding frequency response and other specs, but this isn't essential. Feel free to experiment with whatever mics meet the polar pattern requirements.

M-S uses the center or *mid* mic with the bi-directional mic to achieve stereo. The mid-cardioid mic points directly at the sound source and picks up the direct sound, while the off-axis bi-directional mic picks up the room ambience and reflected sound. With M-S stereo, one channel consists of the center mic signal summed with the side mic signal, while the other channel consists of the center mic signal *minus* the side mic (that is, a phase-inverted copy of the side mic). The center mic has positive polarity and is common to both sides. The left and right sides originate from the same mic, but because the phase is inverted in one channel, collapsing an M-S recording to mono cancels out the left and right sides from the bi-directional mic. This leaves only the positive polarity signal from the center (cardioid) microphone. This significant advantage of M-S recordings insures perfect mono compatibility, without any phase issues.

Setup

As with normal cardioid mic placement, you'll want to aim the cardioid mic directly at the sound source. If you're a fan of close miking, with M-S recording, try moving back a bit farther from the source than you usually would.

Next, place the figure-8 mic so that the two pickup lobes of its pattern are set 90 degrees relative to the cardioid microphone. M-S is a coincident microphone technique, so you want the two mics' diaphragms as close together as possible. Fig. 5.5 shows a Soundelux (now known as "Bock Audio") E250 (bottom) and ELUX 251 (top) set up as an M-S pair. The cardioid E250 points at the sound source (in this case, the camera). The ELUX's pattern selector is set to bi-directional. It's picking up sounds to the left and right, and its side null point aims directly at the sound source/camera.

Figure 5.5 A Soundelux E250 (bottom) and ELUX 251 (top) set up as an M-S pair.

Setting Up the Mixer and Recording

To record this pair, you'll need to route each mic to its own preamp. You'll assign the cardioid mic to a single track in your recording software. For the bi-directional mic, you have three choices:

♦ Split it into two identical, separate record tracks, and invert the polarity of one of the two tracks later.

♦ Record the bi-directional mic to only one track, and use a decoder plug-in to create the sum and difference signals.

♦ Duplicate the single bi-directional mic track after recording, and then invert the polarity of the duplicate track.

In Pro Tools, a flexible option is to record the bi-directional mic to two tracks (labeled *Side+* and *Side-*) simultaneously, and insert a Trim plug-in on the *Side-* track to invert the phase. Fig. 5.6 shows a basic M-S track arrangement in Pro Tools.

In this application, the Trim plug-in is used only for its ability to apply a phase inversion.

Figure 5.6 A basic mid-side track and panning arrangement in Pro Tools. The Trim plug-in reverses the polarity (red arrow) on the cloned *Side-* track.

You'll need to pan the two *Side* tracks hard left and right, then group the *Side* tracks so that any level changes apply equally to both tracks. Raising the level of the *Side* tracks will widen the stereo image; lowering them will decrease it. Being able to adjust the amount of stereo information in the recording after the fact is one of the big advantages of M-S recordings.

Free Decoder Ring Inside!

Although a decoder or hardware matrix box isn't strictly necessary, the downside of simply cloning the side mic track after recording and inverting its polarity is that you can't hear the final result as you position the mics. Instead, you need to complete the recording, duplicate the side mic track, then flip the polarity on the duplicate and start playback. By encoding the sum and difference data from the two mics and recording the result, you can hear how the stereo field will sound before you track.

PAiA Electronics' website provides a hardware M-S matrix box schematic. It is available here: http://www.paia.com/ProdArticles/msdecwork.htm.

For a software solution, Voxengo's free MSED M-S decoder plug-in (Mac AU/VST, Win VST) works very well. You can download it from here: http://www.voxengo.com/downloads/.

M-S is not as common as some other stereo techniques. But with perfect mono compatibility and the ability to adjust the stereo width at mixdown, it is well-suited for broadcast production, live recordings of small ensembles, individual instruments, and small groups of background vocalists on multitrack music sessions.

Blumlein Stereo Pair

While working as an electrical engineer for a Bell Labs subsidiary, and later for EMI, Alan Dower Blumlein (1903–1942) had a profound effect on the fields of telecommunications, TV, radar, and audio recording and reproduction. He developed the first "weighting networks" to compensate for the non-linearity of our ears, designed moving-coil mics, and much more. But probably Blumlein's biggest contribution was that he essentially invented stereo recording and playback. Back in 1931, he also created the Blumlein Stereo Pair, one of the most useful stereo mic techniques ever created (UK patent 394,325; "Improvements In and Relating to Sound-Transmission, Sound-Recording, and Sound-Reproducing Systems").

Blumlein Theory

The Blumlein Pair is a crossed coincident pair of figure-8 velocity (ribbon) mics, each placed 45 degrees off-axis from the sound source, while 90 degrees off-axis from each other—similar to setting up two cardioid mics in an XY stereo configuration (Fig. 5.7).

Figure 5.7 Two RCA 74B Junior Velocity mics configured as a Blumlein Pair.

Placing two figure-8 ("bi-directional") mics in the X position provides a detailed stereo image, while the rear lobes of the two figure-8 mics pick up a significant amount of room tone or ambience and reflections. As with mid-side recording, you'll want to place the mic capsules as close together as possible. Note that the Blumlein technique doesn't require polarity inversion or decoding.

To ensure balanced stereo recordings, a Blumlein Pair utilizes two closely matched figure-8 mics. While the patent specifies a pair of bi-directional ribbon mics like the two RCA 74B Junior Velocity mics shown in the above image, two identical, multi-pattern condensers set for figure-8 polar patterns will work. A factory-matched mic pair is preferable, but you can use any two mics of the same type and model, as long as each offers a bi-directional polar pattern.

If you already own one multi-pattern condenser mic, or one bi-directional ribbon, consider purchasing a second mic of the same make and model so you can exploit the Blumlein Stereo Pair technique.

Using the Blumlein Stereo Pair

To configure your setup for recording using the Blumlein technique, aim one bi-directional mic 45 degrees to one side of the sound source's center; then place the other mic directly above the first, and aim it 45 degrees to the sound source's other side. Make sure each mic's front side points forward, toward the sound source, so that each is in phase. Referring again to Fig. 5.7 above, orient the mics so they're aimed at the sound source (the camera that took the photo is in the sound source location).

Next, route each mic into a separate mic preamp and record each to a separate track, panning the two tracks hard left/right. You can also record each mic to one channel of a stereo track.

As with mid-side recordings, the Blumlein technique picks up a significant amount of room ambience, so you'll need to adjust the mic placement for the desired ratio of direct sound to room ambience (closer to the sound source for more direct sound; farther away to pick up more reflections and reverb).

Despite its age, the Blumlein Pair technique still has significant advantages:

♦ Unlike spaced-pair stereo recordings, phase issues are generally not a problem due to the coincident mic placement.

♦ Because Blumlein Stereo Pairs use amplitude differences, not wave-phase differences, mono compatibility is quite good.

♦ The stereo imaging is realistic—it's very similar to what you hear if you stand where the mics are placed.

 For drum recordings, try setting a Blumlein Pair just behind and above a drummer's head, pointing forward toward the kit. This captures a perspective of the entire kit as the drummer hears it, along with a healthy amount of room reflections.

Blumlein pairs are also suitable for live ensemble recordings, brass sections, and stereo recordings of individual instruments (e.g., acoustic guitars and classical piano)—basically any situation where more ambience is desired than what an XY stereo configuration or close miking provides. So the next time you're looking for a sense of space and ambience in a recording… say hi to Mr. Blumlein.

Don't Forget Omni Mics

Many people will reach for a directional microphone out of habit, or by default because that's all they own. Although cardioid microphones have many advantages, some characteristics—like the proximity effect—can be problematic.

When ultra-close miking provides the sound you want, but you're getting too much bass buildup due to the proximity effect, try an omnidirectional microphone instead. These mics don't have the rear rejection of cardioids, but they also are completely free of proximity effect. While you won't be able to position the mic's back to reject other sound sources or room ambience, often the source-to-ambience ratio will be high enough due to the mic being so close to the source. In this case, you'll be able to reject most unwanted sounds without the added "boom" from the proximity effect.

Key Takeaways

- There are many possible stereo miking techniques, each with advantages and disadvantages.

- A-B (or spaced-pair) stereo miking is very common. It's used more for multitrack productions than live stereo recordings.

- A-B stereo can result in very natural recordings with a wide, deep stereo image. If the stereo sound-field width is of primary importance, then consider A-B stereo a priority. However, mono compatibility is not its strength.

- XY stereo offers better mono compatibility than A-B stereo, and allows capturing more of the source sound and less of the room ambience while still maintaining a good stereo image.

- ORTF gives a wider stereo image than XY miking, although good mono compatibility is not always assured.

- Mid-side stereo provides near-perfect mono compatibility, and allows altering the amount of stereo information (and therefore the image) during the mixing stage.

- The Blumlein Stereo Pair technique provides a detailed stereo image, but also picks up a fair amount of room sound and ambience. Mono compatibility is quite good, and the stereo imaging is realistic.

- Don't forget about mics with an omni response. The lack of proximity effect can be exactly what's needed in some close-miking situations.

Chapter 6

Miking Specific Instruments

Now that we've covered various mic types and miking options, let's look at miking techniques for specific instruments.

Vocals

The vocal is the main element that most listeners connect with. It conveys emotion in addition to the main melody and the lyrical message, so it's crucial to get it right. The book *How to Record and Mix Great Vocals*, another in the Musician's Guide to Home Recording series, covers recording and mixing vocals in detail; here we'll examine the mic-related aspects.

The Mic Is the Singer's Instrument

Some mics flatter some voices more than others. The mic that sounds best with your voice could be a vintage, expensive German mic… or a cheap mic from a garage sale. So record yourself singing through lots of different mics, and choose the one you like best (if one comes close, try using signal processing to help dial in the sound). Remember—the listener won't care what mic you used, as long as they like the vocals.

Mic selection depends mostly on the singer's style and the type of sound you want. Many modern recordings feature lead vocals recorded with a large-diaphragm condenser mic. A trend over the past several years has been to use fairly bright-sounding mics to help a singer cut through a busy mix. However, that same brightness might not be appropriate for a singer with a brighter-sounding voice or a singer who emphasizes sibilants. (Although de-esser plug-ins can tame sibilants somewhat, it's best to get the sound right at the source.)

Here is a list of some popular, high-end vocal mics:

- Neumann M49

- Neumann U47

- Neumann U67

- AKG C12

- Telefunken ELA M 251

The following mics are fine choices for those on a real-world budget:

- AKG C414
- Mojave Audio MA-200
- Studio Projects C1
- Neat King Bee
- Avantone CV-12
- MXL V89
- RØDE NT1

Some very successful records have been made with moving-coil dynamic mics. Classic mic choices include the following:

- Shure SM7B
- Shure SM58
- Sennheiser MD 441
- Electro-Voice RE20
- Electro-Voice RE320

 Don't forget ribbon mics! Before the dawn of condensers, the RCA 44 was a common vocal mic. Modern ribbon mics can work well for vocals—especially when you want a smooth, sweet "crooner" sound.

To Hold or Not to Hold?

Most recording engineers set up a mic stand to hold the mic, thus avoiding noise from the singer handling the mic. But some singers need to hold the mic for the right feel. In that case, as long as you're aware of the potential for noise, and holding the mic gets a better vocal performance, you should do it.

 DPA Microphones' d:facto handheld condenser vocal mics are designed to minimize handling noise. These are excellent, albeit pricey, microphones for vocalists who like to hold their mics.

Mic Technique

It's good to get carried away with your performance, but vocalists need to reserve some concentration for making the audio happy. The most natural, artifact-free, and super-low-cost dynamics control processor is great mic technique—moving closer for more intimate sections and farther away when singing more forcefully.

Mic Positioning

When using a mic stand, position the singer so their mouth is six to eight inches away from the mic, with a pop screen half-way between the two (Fig. 6.1). Moving closer can give a thicker, fuller sound from the proximity effect, as well as a more intimate sound. If proximity effect is a problem, move the singer back about a foot from the mic.

Figure 6.1 A pop filter will help reduce plosives, thus minimizing the need to edit them out when mixing.

When recording vocals close to a mic, use a good pop filter or screen to minimize plosives such as p-pops. These screens also establish a minimum working distance that keeps overly-enthusiastic vocalists from spraying into your expensive mic. To reduce room-ambience pickup, use a baffle or other audio isolation tool (such as the sE Electronics Reflexion Filter or Primacoustic VoxGuard) on the sides and behind the mic.

Background Vocals

If the singers are still working out the parts, or if you think you might want to process (or correct) the vocal tracks individually, overdubbing individual background vocalists one track at a time is a good approach.

For a more blended sound, position the singers around a single omni-directional mic as they sing their backing parts simultaneously (Fig. 6.2). However, the singers need to be well rehearsed, because any mistakes will require re-recording the part.

Before recording, move individual singers closer to or farther from the mic until you hear the right balance. You can then double-track, or multitrack and stack parts, and pan parts in stereo to give the impression of a larger group of singers. This also gives thicker-sounding parts.

Figure 6.2 Background vocals usually work better in a mix if they're farther away from the mic than the lead vocal.

While mics used for lead vocals can also be effective on background vocals, consider making two changes from the lead-vocal miking approach:

♦ Substitute a different mic so the backing vocals have a different sound quality from the lead vocal. This is particularly helpful if the same person is singing all the tracks. Example: If the lead vocal sound is low and dark, try a brighter-sounding mic for the background vocals.

♦ Remember—distance equals depth. Physically placing the background vocalists anywhere from one to three feet further away from the mic than the lead singer gives a different, less upfront sound that will help differentiate the lead and background vocals when you mix.

Guitar Amps

Recording an amp isn't as simple as pointing a mic at a speaker and clicking the record button. The classic approach is to place a dynamic mic right up next to the grille cloth, but placement matters—moving the mic just a few inches can vary the sound considerably, for better or for worse.

Selecting the Right Mic

Some mics, like the industry-standard Shure SM57, have a presence peak in the 5 to 7 kHz range that can help a guitar cut through a mix and sound more articulated. However, this works against you if the amp sounds too edgy or bright. A warmer-sounding ribbon mic that lacks a presence peak may be a better choice.

Moving-coil dynamic mics are popular for miking amps, because they're low-cost, rugged, and can handle high sound pressure levels. However, ribbon and condenser mics offer different sonic characteristics. Ribbons tend toward a warm, natural sound that captures the initial note attack more accurately than moving-coil dynamic microphones. Condensers usually sound brighter and clearer.

While the SM57 is probably the most popular guitar amp mic, other commonly used dynamic models include the following:

♦ Audix i5

♦ Sennheiser e 609

♦ Sennheiser MD 421

♦ Audio-Technica ATM650

♦ Electro-Voice RE20

Ribbon mics can work well for amps, and their figure-8 response opens up options when miking two cabs. Although a ribbon's high-frequency response is somewhat subdued compared to many condenser mics, this characteristic can minimize harshness—yet its condenser-like, quick transient response still provides a clear, detailed sound.

Here are some common ribbon choices for amps:

♦ Beyerdynamic M 160

♦ Royer R-121

♦ MXL R144

♦ Cascade Fat Head II

♦ AEA R92

For more sparkle and chime, consider a large-diaphragm condenser mic:

♦ Neumann U67

♦ AKG C414 EB

♦ Neat King Bee

♦ Mojave MA-201

♦ Mojave MA-300

♦ Neumann TLM102

♦ Bock Audio 195

♦ Earthworks' SR Series

 Using mics with different characteristic sounds on different tracks can provide more distinct and individual tones, so they stand out more easily in the mix. *Example*: To emphasize a jangly guitar track's attack or treble, try a brighter condenser mic. For a smoother sound, try a ribbon.

Levels and Tones

You don't need excessively loud amp levels to capture a great electric guitar sound. As long as the amp creates a good tone, all you need to do is capture it. Many classic guitar recordings used small, low-wattage tube amps that could break up and distort at lower volume levels. Turn up your amp only as loud as needed to obtain the desired tone.

 Distortion is tricky in the studio. When mixed with other tracks, too much distortion can clutter the mix and make the guitar part less distinct. Experiment with lower gain levels on your amp and stompboxes, if possible in context with other tracks.

Dial up the guitar, effects, and amp sound so that it sounds good in the room. If it doesn't sound right coming out of the amp, you probably won't be able to fix it in the mix. Make sure you're happy with the sound before continuing.

Also, don't "slam" your recording levels. Average signal levels should reach about –18 to –15 dB below 0 on your meters. If your meters aren't calibrated, but just have LEDs, the average signal should hit the middle of their range. Peaks can exceed that level occasionally, but avoid running into the red and the risk of clipping your recording.

High-Pass Filters

A mic, preamp, or audio interface's high-pass filter can minimize sounds below the guitar's range that don't contribute to the guitar sound. These frequencies can cloud the bottom end when combined with other instruments. High-pass filters are common on condenser microphones, but rare with ribbon and dynamic mics. However, while mixing, applying a high-pass filter plug-in in your recording software set around 80 to 100 Hz, with the steepest possible slope (Fig. 6.3), can clean up a guitar amp's low end.

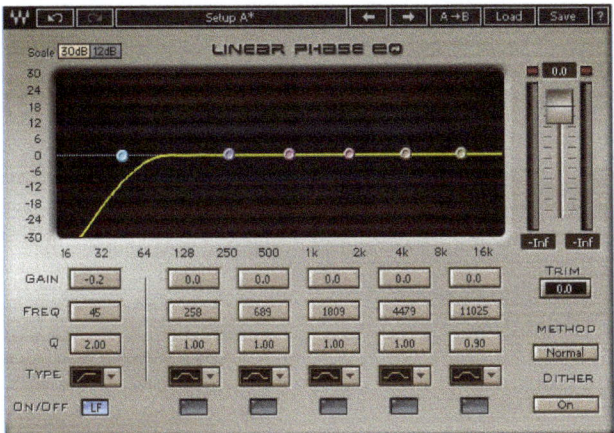

Figure 6.3 This high-pass filter from Waves minimizes unneeded low-frequency energy. The cutoff frequency could likely go higher without affecting the guitar tone.

Miking the Speaker

Although some amps have a direct out, you'll likely be miking the amp's speaker. Here's what you need to know.

Locating the Speaker

A speaker sounds different if you point a mic toward its center, off-center, or at its edge. Fairly transparent grille cloths make it easy to see where the speaker is. Other grilles are almost opaque, and spotting the speaker is trickier. This matters when miking, because speakers aren't always centered.

Shining a flashlight through the grille, looking through the grille up close, or viewing it from an angle may help. Some amps have removable grilles, and pulling the grille off can offer better speaker access—but be careful! The grille protects the speaker from damage, so use caution.

Multiple Speakers

When a cabinet has more than one speaker (like a 4 x 12 cabinet with four 12-inch speakers), you can experiment with placing the mic on each speaker. One may sound better than the others.

Some newer amps are more like PA systems designed for flat response and have both a low-frequency and a high-frequency driver. The most-balanced mic placement is usually at the *acoustic axis point,* which is midpoint between the low-frequency and high-frequency driver. The mic needs to be far enough away from the amp (e.g., one meter or three feet) to pick up both speaker sounds without favoring one over the other. If you wish to favor one, aim the mic accordingly.

A few amps go beyond two drivers. The Line 6 Firehawk 1500 is a stereo unit with six speakers. It can deliver a flat response (e.g., to serve as a keyboard amp) as well as guitar tones. Recording an amp like this is more challenging. The simplest solution is to place a mic far enough away to pick up the combined acoustical output of all speakers. Another option is to use separate mics for the different speaker types, or use a speaker cabinet-emulated direct output (if present) and use mics more for picking up the room ambience.

Microphone Positioning and the Horizontal Line

How you position the mic can change the basic sound without using different mics. Imagine a horizontal line that starts at the speaker's outer side edge (called the *surround*), and goes through the speaker's center over the dust cap. Moving your mic sideways across this horizontal line while pointing at the speaker has a dramatic effect on the sound. Pointing the mic toward the speaker's center, at a 90 degree angle compared to the grille cloth, provides the brightest tone (Fig. 6.4).

Figure 6.4 This mic position captures the brightest amp tone.

Moving the mic toward the speaker's outer edge creates a mellower and progressively darker timbre. (Fig. 6.5).

Figure 6.5 The mid-cone position is slightly darker and warmer sounding.

Placing the microphone at the speaker's far edge and aiming directly at the speaker surround creates the darkest sound (Fig. 6.6).

Figure 6.6 The speaker's edge deemphasizes the high frequencies.

On- and Off-Axis Mic Placement

The mic's angle in relation to the speaker also matters. *On-axis* positioning points the mic directly at the speaker, with the mic at a 90 degree angle relative to the grille cloth. *Off-axis* positioning changes the mic angle to 45 degrees (Fig. 6.7). While on-axis placement is common, off-axis placements provide additional tones.

Figure 6.7 Here the microphone is off-axis, at a 45 degree angle relative to the speaker grille.

Placing the mic by the speaker's side edge, but angling it in 45 degrees so it points toward the center dust cap, provides some of the advantages of both positions—the brightness of the center position, with the warmth of the edge-of-speaker position.

Distance Gives Depth

Placing the mic within an inch or two of the grille cloth is common (Fig. 6.8). This gives a dry, "in your face" sound that can be perfect for some contexts. However, you always hear an amp in a room. To create a sense of space, you can add short delays or reverb during mixdown to maintain the close-miking sound, but also create some ambience.

Figure 6.8 Close-miking captures the amp sound, but does so at the expense of the room sound.

When playing through an amp, the room becomes part of the sound you hear. So experiment with more distant mic placements for a more realistic, nuanced sound that captures more ambience, room reflections, and reverb. Positioning a mic eight inches (or more) away from the grille (Fig. 6.9) gives a different sound compared to close-miking, even if the mic points to the same place.

Figure 6.9 Distant mic placements can provide a more ambient sound.

Remember that you can't remove ambience captured as part of a recording. If this is a concern, try recording two mics, one close-miking the amp and the other capturing the room sound. Capturing these elements on separate tracks allows you can blend them to taste at mixdown. Alternatively, you can use three mics—one for close-miking the amp and the other two set up as stereo room mics.

Mic Placement Training for Amps

Because mic positioning and mic type is crucial to getting a good sound, you'll want to gather as much experience as possible with miking amps. However, you may not have the option to do so, or have enough gear to become experienced on your own before walking into a studio.

Fortunately, today's amp simulation software typically includes miking options with varying degrees of sophistication. Amp sims typically allow for choosing different mics, mixing them together, changing the angle, and altering the position on the speaker (Fig. 6-10).

Figure 6.10   IK Multimedia's AmpliTube 4 offers multiple miking options for two speakers. This shows a condenser mic at the cone and a dynamic mic at the edge.

Although an amp sim isn't the real thing, it's like learning the basics of flying through a flight simulator, and it will give you an idea of what to expect when miking a physical amp.

Acoustic Guitar

A great acoustic guitar will simplify recording a great acoustic guitar sound. Replace the strings a day or two in advance of a session. You want them to be fresh but also to have had time to stretch and settle in so they'll stay in tune. Applying some Finger-ease string lubricant or a little talcum powder can help reduce finger squeaks.

Small-diaphragm condenser mics are popular for acoustic guitar, but large-diaphragm condensers are also applicable. Even ribbon mics are suitable, particularly to tame an overly bright guitar.

Typical condenser options include the following:

- MXL 603
- Audio-Technica AT4041
- Audio-Technica PRO 37
- Oktava MK-012
- Shure SM81
- Studio Projects B1

Here are some common ribbon mics:

- Royer R-101
- Beyerdynamic M 160

 For classical guitar players with a light touch, try the Earthworks QTC Series. These mics are designed specifically for quiet, low-level sound sources.

Mic Positioning

Recording with a single mic eliminates the potential phase cancellation of stereo miking. A common placement will position the mic about 6 to 12 inches directly in front of the 14th fret, angled in slightly so it points toward the fingerboard between the body's edge and the sound hole (Fig. 6.11).

Figure 6.11 Using a single mic prevents phase cancellation, but the tradeoff is a lack of stereo image.

For stereo, add a second mic about the same distance away, but placed over the lower bout (the lower curve of the body) or just behind the bridge (Fig. 6.12).

Figure 6.12 When miking in stereo, sum the signals in mono to make sure there aren't phase-cancellation issues. If so, try moving one mic slightly to see if that provides a solution.

Another option is a classic XY stereo pair, placed about 12 to 18 inches in front of the guitar. Center this stereo pair so that it's almost directly in front of the sound hole (Fig. 6.13). Pointing a mic directly at the hole usually isn't recommended because it can sound muddy and boomy, but in the XY pair's 90-degree angle, the two mics will point to either side of the sound hole instead of directly into it.

Figure 6.13 XY miking can give a full-sounding stereo recording.

Rotary Speaker Cabinet

The Leslie speaker was invented by Don Leslie and made its commercial debut in 1941. It consists of a cabinet containing a power amplifier, a pair of speakers, and a system of motors and pulleys to spin acoustic baffles that redirect the sound from the two speaker elements. This adds considerable complexity, motion, and interest to any sound the system processes.

While the Leslie speaker is associated with Hammond organs, it has been used successfully with sound sources like guitars, vocals, drums, and other pre-recorded tracks via re-amping.

What's Inside the Box

Capturing the sound of this electro-mechanical beast requires knowing how it works. A typical Leslie cabinet, such as the Model 142 (Fig. 6.14), has three sections:

♦ The cabinet's center houses the crossover and two speakers—a downward-firing woofer or *drum* and an upward-firing, high-frequency *compression driver* or *horn*. The crossover splits the audio into two frequency bands. Signal below 800 Hz goes to the woofer, and signal above 800 Hz goes to the compression driver in the cabinet's upper section.

   While there appears to be two opposing horns, the high frequencies from the compression driver go through only one of them. The second one is closed off, and serves only as a counterweight so that the main horn remains balanced as it spins.

♦ The cabinet's bottom section contains a curved "ramp" or scoop that redirects the sound from the downward-firing, low-frequency, rotating woofer out of the cabinet's lower front, sides, and rear. This section also houses the 40 W tube power amplifier. While some later models may have a solid-state amp, or rely on an external amplifier, the classic models have a tube amp in the lower right-hand corner when viewed from the front (as in Fig. 6.14). This amp can be a source of noise.

♦ Most Leslie speakers can also switch the motor from a slow rotary speed (sometimes called *chorale*) to a fast (*tremolo*) speed. The solenoid switches that control the motor speed are located near the power amplifier, and produce a loud "click" or "clunk" sound when changing speeds. Avoid that side of the cabinet when placing your mics (especially the low-frequency mic for the woofer).

Figure 6.14 A Leslie Model 142, seen from the front.

Rotating Speaker Basics

Changing the rotation speed alters the sound in novel ways, due in part to the different mass of the rotary horn and drum elements. The larger, heavier drum takes more time to speed up or slow down than the horn. The interaction of the sound from the two speakers—phase shifts, sound reflections off the cabinetry, Doppler-shift pitch changes, and amplitude (tremolo) fluctuations—make for an extremely complex sound, especially when the motor speed changes and the drum and horn transition to the new setting.

Tech Talk: How It Works

The upper horn's effect modulates frequency. As the horn spins toward the listener's position, the sound's pitch rises slightly. When spinning away, the pitch drops slightly. The sound of the Doppler shift changes, depending upon the horn's speed. The horn rotates at about 50 RPM in Chorale mode, and around 400 RPM on the Tremolo setting (the drum rotates at about 40 RPM in Chorale, and 340 RPM with Tremolo; it can take five or six seconds to transition speeds). Due to the lower frequency, the drum's overall effect is mostly an up-and-down volume fluctuation as the drum turns toward and then away from the listener. However, there are elements of both amplitude and frequency modulation with both drivers; the amount of each you capture on your recording depends upon how closely you place the microphone(s) to the cabinet, as well as where you place them.

How Many Mics, and Where to Put Them

Ask ten engineers where to place a mic, and you'll get ten different answers. However, some characteristics are unique to a Leslie speaker.

Because the cabinet's two speakers reproduce different frequencies, ideally you need at least two mics. If you *have* to use a single mic, mic the cabinet from the side or rear with the mic about six feet back from the cabinet, three feet above the floor, and aimed straight at the cabinet. Aim or lower it farther toward the drum for more low frequencies; aim or raise it more toward the horn for more high end.

With two mics, one mic can capture the horn while the other records the drum (Fig. 6.15). Record each mic to its own track so you can adjust the balance in the mix.

Figure 6.15 Using two mics provides more flexibility than a single mic. Here a Cascade Fat Head II (top) aims toward the horn, while an Electro-Voice RE320 (bottom) aims at the drum.

The Leslie's front and sides are permanent fixtures, but you can remove the back and mic the speaker from the rear. The sound will be slightly more open and less diffused because the sound isn't passing through the cabinet's louvers, but the downside is that this setup will be more prone to picking up pulley and mechanical noises.

Moving the microphones farther away from the cabinet (Fig. 6.16) creates a smoother sound with more room ambience and less volume fluctuations, while close miking accentuates the tremolo effect—especially at high speeds.

Figure 6.16 Here, the mics have been moved back about three feet. Moving them even farther away can also be effective.

Miking the Horns

Using two mics on the horn for stereo can give a dramatic effect. The rotating horn's high-frequency response doesn't reach much past 7 kHz, so there's no need for mics with extended high-frequency response. Small-diaphragm dynamic mics (like the Shure SM57, Audix i5, AKG D1000E) or ribbon mics (like the Cascade Fat Head II) are common choices instead of condenser mics.

Placing the two horn mics on the two opposing sides of the cabinet gives good results. Placing one mic at each of the rear corners, and angling them in slightly towards the horn, also works. Another common technique is to place one mic at the rear and the other on one side (Fig. 6.17).

Figure 6.17   A pair of Cascade Fat Head II's are placed fairly close, one at the rear and one on the side, to capture the Leslie horn in stereo. The pop filter in front of the rear mic (left) helps protect it from wind gusts.

Miking the Drum

A single microphone is sufficient to capture the drum. Large-diaphragm dynamic mics like the Sennheiser MD 421, the Electro-Voice RE20 and RE320 (pictured in Fig. 6.18), and the Audio-Technica ATM250 work well. Even with fairly distant placement for the horn, some engineers will close-mic the low frequency drum.

Figure 6.18   An Electro-Voice RE320 aims at the low-frequency drum.

With this technique, consider the following:

♦ A high-pass filter for the horn mics set around 800 Hz reduces low-frequency bleed from the drum that's picked up due to the more distant placement.

♦ Some engineers keep the drum mic close while moving the horn mics farther back. If you have a significant distance between the mics, you may need to nudge one of the tracks forward or backward on the timeline to maintain phase and time alignment.

 For more on maintaining phase alignment, see the discussion below on recording bass.

Cautions

The power amp in many Leslie speakers is in the enclosure's lower-left side (viewed from the back), and is notorious for producing electronic interference and hum. The components also generate mechanical noises when changing the speed switch. Positioning the mic on the side of the drum opposite the amp will minimize this noise.

The drum can generate wind noise, so use a foam windscreen over the drum mic. Usually placement alone can minimize wind noise issues, but if the noise persists, insert a vocal pop filter between the mic and the speaker.

Areas close to the cabinet openings and vents are prone to wind gusts from the spinning horn and drum, and a windblast can destroy a fragile ribbon element. Place any ribbon mic back a bit from the cabinet to avoid direct air blasts, or use a pop filter for protection.

It's virtually impossible to eliminate the sonic artifacts from pops, clicks, hum, and wind noise, but you can minimize them so they're less obtrusive. In fact, it's entirely possible that the absence of that noise can cue listeners to whether they're hearing a "real" Leslie or a good simulation.

Electric and Synth Bass

The importance of bass in modern recordings can't be overstated. Bass parts range from the deep boom of a dub bass to the bright percussive attack of slap bass, so it's important to decide what type of sound you want and choose the gear that can deliver it.

The sound starts at, and is only as good as, the bass itself and how it's played.

♦ Check the bass for accurate intonation and setup.

♦ Use fresh, roundwound strings for a bright, full-frequency modern sound.

- Use flatwound strings for vintage Motown and 60s rock tones.

- Different bass models have characteristic sounds. For the sound of a particular bass, use that model bass, if possible.

- Use the appropriate playing technique (pick, fingers, thumb, or slapping). Fingers give a softer, old-school bass sound while a pick emphasizes transients and can help the bass stand out in a busy rock mix.

- Set the pickup distance to the strings closer for more level, or farther away for more sustain.

- The vibrations from bass amps can affect other gear in the studio. Check for rattles or loose hardware before hitting record.

Recording: The Four Main Methods

Following are the four main recording methods for electric bass.

Direct (also known as DI or Direct Input)

If your recorder or audio interface has a high-Z (high-impedance) input (Fig. 6.19), you can plug the output from your bass or effects pedals directly into it and record with no additional hardware. Some processing with compression and/or EQ plug-ins can give solid bass tones.

Figure 6.19 A typical audio interface high-impedance input.

If your audio interface doesn't have a high-Z input, a *direct box* can be used to convert the bass's output impedance so it's compatible with an audio interface or mixing console's mic or line input. A direct box will have at least an input and output. You can plug your bass into a direct box input, and then patch the Direct Box's line output to your mixer or audio interface. There may also be a *through* or *pass-through* output that carries your bass signal, and is suitable for plugging into a bass amp.

Several hardware units also emulate amp and cabinet sounds, like the Line 6 Helix, SansAmp VT Bass, DigiTech BP355, and BOSS GT-10B. Depending upon the unit's output level and impedance, it may be able to plug into a high-Z input, standard line input, or both—check your product's manual. Like a physical amp (and unlike software), the processed sound is "baked into" the recorded track.

Direct with Amp Simulation Software

After plugging your bass into a high-Z input or direct box, you can emulate bass amps and cabinets in your recording program with amp simulation plug-ins like Native Instruments Guitar Rig, Line 6 Helix Native, Waves GTR, and IK Multimedia Ampeg SVX (Fig. 6.20). These all include bass amps and cabinets as well as ones for guitar.

Your audio is always recorded on the track with no processing; you hear the amp sim's effect on playback. To monitor this sound as you play, you may need to enable a feature in your host program called either *input echo* or *input monitor*. Because you will be listening to the signal after it passes through the computer, this may cause a short delay called *latency*. Some programs have low-latency modes for monitoring the processed audio while you record. Also, some audio interfaces have a *zero-latency* mode that monitors the signal as it goes into the audio interface. This eliminates latency, but you won't hear the effect of the amp sim.

Figure 6.20 Software amp simulation plug-ins such as IK Multimedia's Ampeg SVX can give your bass the sound and character of a miked amp. This preset is modeling a vintage Ampeg B-15.

Most amp sims allow for *parallel processing* so you can combine the unprocessed and processed bass sound. If not, you can record the bass direct, copy the direct track or send it to a bus, process the copy or bus with the amp sim, and then mix the two together for interesting combination tones.

Miked Bass Amp

The basic principles are like those covered previously on miking guitar amps; however, bass usually covers a wider range of frequencies. Unlike guitar amp speaker cabinets, which typically reproduce frequencies from 100 Hz to 6 kHz, bass amps can extend considerably lower and higher. Many bass rigs even have high-frequency tweeters to complement their low-frequency bass drivers, and the trend for many years has been toward *full-range* (i.e., flat frequency response) bass amplification systems. These not only reproduce the fundamentals faithfully, but also the harmonics and attack that are part of modern bass playing—especially with slap and pop playing techniques.

As with miking a guitar amp, moving the mic toward the speaker's edge gives a tighter sound. Pointing directly at the speaker's center dust cap produces the brightest and fullest sound (Fig. 6.21).

Figure 6.21 As with guitar, the speaker's center generates more high frequencies than the edges. Close, on-axis miking tends to give the "roundest" sound.

Some bass amp cabinets contain small tweeters. Depending upon their location, moving the mic toward the edge of the cone closest to the tweeter can give a brighter sound than the center dust-cap position (Fig. 6.22).

Figure 6.22 Placing the mic at the edge of the speaker cone normally gives a tighter, darker sound. However the tweeter in the upper right corner adds brightness when positioning the mic near this location.

Alternately, adding a second mic for the tweeter can create additional brightness. By recording it to a separate track, you can adjust the woofer/tweeter balance when mixing.

Mics that work well on bass cabinets have good low-frequency response, along with enough mids and highs to provide balance. Here are some common "big studio" choices for large-diaphragm, dynamic mics:

♦ Electro-Voice RE20

♦ Sennheiser MD 421

♦ Audix D6

♦ Heil PR 40

♦ AKG D112

♦ Audio-Technica ATM250

♦ Electro-Voice RE320

Large-diaphragm condenser mics can also work well. Ribbon mics work in some situations, but the bass boost from their inherent proximity effect can be too much with bass guitar.

♦ Neumann U47 FET

♦ Neumann TLM102

♦ RØDE NTK

 Experiment with moving the mic back a bit farther from the bass cabinet than you would with guitar. Bass wavelengths are considerably longer, and placing the mic where the waveform has developed more may provide a better sound.

Bi-Amplified Bass Cabinets

With *bi-amplified* bass rigs that use separate speaker enclosures for the low and high frequencies, consider using separate mics optimized for each cabinet—e.g., a dynamic mic such as the RE20 for the low frequency cab, and a condenser for the high frequency one. Recording each mic to its own track retains control over their relative balance when mixing.

Combinations of Different Techniques

The bass guitar produces low frequencies that certain gear has difficulty reproducing. Combining some of the above techniques can help create an even, full sound.

A common combination is direct sound with a miked amp, each recorded to its own track. The direct sound provides a consistent response, while the amp adds character. Use the best aspects of each to build your final bass sound during mixdown.

Amp sims that allow for parallel processing can combine unprocessed sound and simulated amp sound inside your computer. This can provide novel effects, like layering tempo-synched filtering changes that lock the bass sound to tempo.

Phase Issues with Multiple Techniques

A time difference between two parallel paths can cause phase-related issues. For example, a direct path will have no delay, but if the mic on an amp is a few feet away, there will be a few milliseconds of delay in its signal. Although you won't hear an actual echo, the low frequencies that are essential to a good bass recording may cancel partially and weaken the sound. To compensate, zoom in on your software's waveform display at the start of the first bass note to see how the tracks align (Fig. 6.23)

Figure 6.23 This illustration shows three bass tracks recorded simultaneously through an amp, an amp sim pedal, and a DI box. The DI track (bottom) is out of phase with the other two tracks because it's early compared to the amp and amp sim tracks.

It's important to align these tracks so that the waveform peaks and troughs correspond. Either move any miked sounds earlier, or move the DI sound later, so that all the audio peaks and dips match (Fig. 6.24).

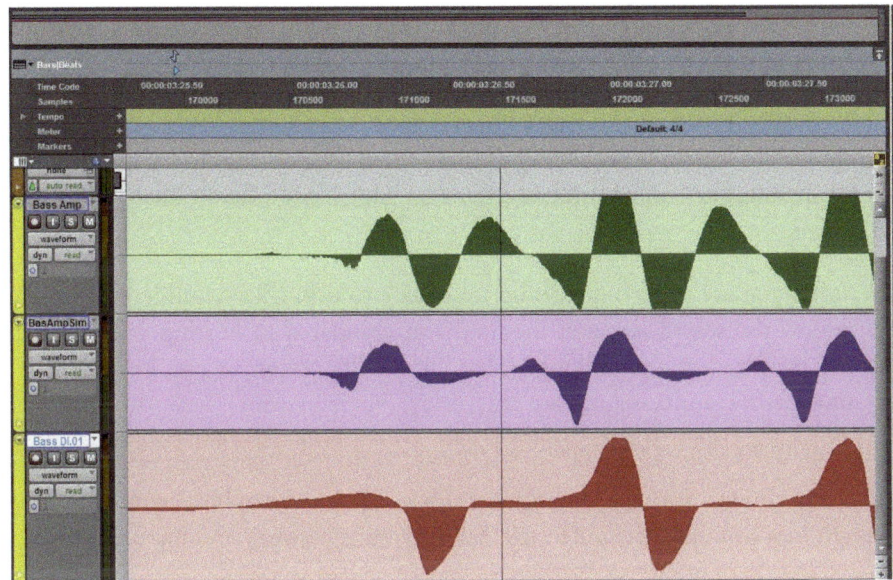

Figure 6.24 The same bass tracks after moving the DI track a bit later to alignment it with the other tracks.

Choosing the Right Method

Any of the above techniques can provide a good bass sound, but it's important to choose the option that delivers the *right* bass sound.

Bass Amp Pros and Cons

If you have a great-sounding bass amp, miking it makes sense. A physical amp can produce a growl and depth that amp sims may not be able to duplicate. Furthermore, you can experiment easily with different miking techniques that a sim may not offer.

The downside of physical devices is frequency-response irregularities. Sometimes this adds character, but it may make some notes more prominent than others. Also, tube amps require maintenance. Tubes don't just break one day; they get "soft" over months—and like strings, you have to decide when to replace them. As mentioned previously, the vibrations from amps can cause other objects in a room to vibrate or rattle. You'll need to make sure your mics don't pick up these sounds.

Finally, you can't make major changes to the amp tone after recording it. Although there's a lot to be said for committing to a part, it's prudent to record a direct track as well as a miked track. Then if you need to make changes, you can re-amp the sound on playback through different amp settings.

Amp Sim Pros and Cons

If noise or neighbors are a factor, going direct and using an amp sim allows good tone at low volume levels. Heavy guitar amp distortion is a challenge for amp sims to reproduce, but sims are convincing with bass because it doesn't generate the same kind of high-frequency distortion. It's also impractical to have multiple bass cabinets sitting around, but an amp sim can offer several different types as well as effects.

Another advantage is that you're always recording the bass's direct sound, as it sounds when plugged into your audio interface's high-Z input. The amp sim is always *processing* the bass sound (unless you render or bounce the track with the sim to include the sim's sound in the audio file). This lets you change the bass sound at any time, even when mixing.

The main disadvantage is that unless you are using a fast audio interface protocol (like Thunderbolt), there will be some delay when listening to the sound through the amp sim compared to hearing your playing direct. This doesn't bother some people, but delays longer than about 20 ms or so can be annoying. Interfaces with a zero-latency monitoring option eliminate this delay, but you won't hear the amp sim sound as you play your part.

Another disadvantage is that the amp sim is not really moving air from inside a wooden box, with a mic that captures the moving air. Although in a mix it's difficult to tell the difference between an amp and an amp sim, it's not impossible.

Pro and Cons of Direct Boxes

A direct box captures the bass sound without any of the anomalies caused by mics or amps. If it has an output transformer, that can even add a subtle processing that enhances the sound. However, the inherent sound may be too dry and clinical for some contexts, although re-amping or using an amp sim solves that issue. In any case, it's a good idea to record the direct bass sound—no bass player has ever been known to say, "I never should have recorded that direct bass sound as a safety."

Synth Bass

The usual procedure is to record synth bass directly into a mixer or audio interface input. However, splitting the signal to record through an amp as well can provide growl and character, which is often a better option. As with bass, you can nudge the miked sound a bit earlier during mixdown to avoid phase issues with the direct source.

Drums

Miking a drum kit is one of the most challenging recording tasks. A drum kit may seem like a complete instrument, but it's actually comprised of several individual instruments—kick, snare, toms, cymbals, and hi-hat. Each has its own sound and miking needs; a great kick-drum mic won't be a great snare mic. And due to their close proximity, each drum sound will *bleed into* (get picked up by) mics intended to capture other parts of the kit. (Let's not even consider what would happen if a drummer hit a mic instead of a drum....)

Again—the Source Matters

While drum selection, maintenance, and tuning are beyond the scope of this book, great-sounding drum recordings require a great-sounding drum kit. You don't have to use the most expensive kit, but it should be in good repair, with properly tuned drum heads. Not all drummers keep their drums in peak condition, so hiring a drum tech to set up the kit prior to a recording session can translate into much higher-quality drum recordings.

Kit Size

Larger drum kits usually require a more complex mic setup to cover the sound properly. For example, if it includes a double kick drum, one kick drum mic won't be enough. While a drummer may need a huge rig with lots of extras for live performance, less is often more in the studio—like a stripped-down, four- or five-piece kit (Fig. 6.25). Use this for the main drum parts, and overdub roto-toms, tambourine, wood block, gong, and other parts later, on separate tracks. This gives more mixing options and better track separation—and it doesn't require as many mics and input channels while tracking.

Figure 6.25 A basic five-piece drum kit in the studio, ready for miking.

General Approaches

Although many options are available when miking a drum set, the process generally falls into one of two broad categories:

♦ Capture the kit's overall sound with a stereo pair of overhead mics, and then fill in the sound by close-miking specific drums as needed.

♦ Capture the individual kit elements with close mics, and then fill in the sound with overheads and room mics.

These approaches may seem similar, because they combine overhead and close mics. However, the philosophical difference is considerable. Emphasizing the close mics gives a more detailed, precise sound, whereas relying primarily on the overheads captures more ambience and leans toward a live-performance vibe. Neither approach is right or wrong, although each does have some advantages over the other.

♦ It's simpler from an equipment standpoint to use overheads to capture the main drum sound. This requires fewer mics, but places greater demands on the drummer's ability to control the dynamics and relative balance of the kit elements. It also requires a room that sounds good (which usually, but not always, translates into having good acoustics), because you're recording the room almost as much as the drums.

♦ The close-mic approach, with each mic recorded to a separate track, provides lots of control. It's also optimum if you plan to use drum-replacement software (which works best on individual sounds) to change or augment some of the kit's elements.

Basic Miking Configurations

There's a dizzying array of different drum miking techniques, from the simple to the highly complex. Before getting into specific mic choices for specific drums, let's zoom out a bit and look at potential miking options—from one or two mics all the way up to ten or more. After finding the option that's most applicable to your situation, then you can narrow down which specific mics you want to try.

The basic, time-tested setup for drums uses four mics: kick, snare, and two overhead mics that capture the overall kit sound. From there, you can either subtract or add mics to create other configurations.

The number of computer audio interface input channels will be a limitation for many people. Most inexpensive audio interfaces offer only two mic inputs, but for tracking drums you'll need four or more. An eight-channel interface allows tracking stereo overheads, snare, kick, and three tom mics (or two toms along with a room, ride, or hi-hat mic). For tracking the equivalent of a live session with a band or rhythm section, you'll want at least sixteen inputs.

Note that some audio interfaces are expandable. For more information on audio interfaces, see the companion book *How to Choose and Use Audio Interfaces* in the Musician's Guide to Home Recording series.

Two-Channel Interface with One or Two Mics

While it's possible to capture a great drum sound with only one or two mics, it's very difficult. Not only does this require exceptional attention to mic placement and a great-sounding room, but it also needs a drummer who can balance the kit sound while playing. Below are some mic configuration options.

One Mic

For this technique, you'll need to wander around the room until you find the spot with the best sound; then put the mic there. Try about six feet (two meters) in front of the kit, at about the same height as the drummer's chest, and aim the mic toward the drummer.

Another option is an overhead placement three to four feet above the snare drum head, aimed straight down at the kit somewhere between the snare and rack toms (Fig. 6.26). Cardioid condenser mics (such as the Audio-Technica AT4041 shown in the photo) are the traditional choice for overhead mics, although ribbon models (such as the beyerdynamic M 160) also work well.

Two Mics

Using two mics is still limiting, but it's better than a single mic. Here are three options:

♦ Use one overhead mic and one mic on the kick drum (hey, it worked on the early Beatles records). Place the overhead mic as described for the one-mic option, and position the other mic on the kick as described later in the section on kick drum mics. While this is a solid two-drum mic setup, it's essentially a mono configuration; you won't have stereo tom fills unless you overdub those parts later.

♦ Use one mic on kick and one on snare, if your music relies primarily on those two elements. You'll usually sacrifice clarity with the toms and cymbals unless you overdub those parts later.

♦ Use an overhead mic as described, then add a second mic (Fig. 6.26, lower right) somewhat higher than the floor tom and aimed across the kit at the snare and hi-hat. This can give a reasonably balanced sound.

Figure 6.26 Basic overhead mic placement can be as simple as one mic centered over the kit and aimed down at the snare or kick pedal. Add a second mic (toward the lower right in the image), higher than the floor tom and aimed across the kit at the snare and hi-hat for a variation on the Glyn Johns stereo overhead setup.

Blumlein, ORTF, XY, or AB/spaced stereo mic pairs can work with a two-channel interface and capture a kit's sound, but each relies heavily on the quality of the kit and the room acoustics. None of them will create a big-sounding kick drum without additional help.

Four-Channel Interface with Four Mics

A four-channel interface opens up many more options. The most straightforward configuration is a kick mic, snare mic, and stereo pair of overhead mics. With careful placement, you can generally capture a good balance of the toms and cymbals with the overheads. Move the overhead mics out toward the cymbals, or farther in toward the toms, to achieve the desired balance.

The well-known British recording engineer Glyn Johns pioneered what is probably the most often-discussed four-mic technique in history:

♦ Place one mic overhead, about three to four feet higher than the snare; aim it down between the rack toms, and point it toward the kick pedal.

♦ Place the second overhead mic much lower—only about six inches above the top of the floor tom, and aimed across the top of it toward the snare and hi-hat. An alternative is to raise this *side* mic a little higher and back behind the kit a bit more, but regardless of the exact height and location you choose, both overhead mics must be equidistant from the center of the snare drum's top head. Use measuring tape or a few feet of string or mic cable to check. This centers the snare in the stereo image, and avoids phase issues. Pan the center overhead to about 3 o'clock and the tom side to about 9 o'clock; adjust to taste from there.

♦ Use another mic on the kick, as described later.

♦ The fourth mic goes on the snare. You have a couple options: The traditional approach is over the snare's top edge and angled down (Fig. 6.27).

Figure 6.27 The usual approach to snare drum mic positioning is from the top, just past the rim and angled downward.

Miking the top edge gives a solid attack, but it may be light on the actual snare rattles. It's often better to mic the snare shell's side (Fig. 6.28).

Figure 6.28   Miking the snare shell's side can offer a solid, balanced snare sound. The curved metal piece behind the snare mic is an sE Instrument Reflexion Filter (IRF), which helps reduce hi-hat bleed reaching the snare mic's back side.

Miking the shell side balances the note attack, shell resonance and "body," as well as the snare rattle, without resorting to tricks like separate top and bottom snare mics.

Avoid aiming the snare mic at a drum's vent hole, or a blast of air will hit the mic with each snare hit.

Record each mic on a separate track in your recording software so you can adjust their relative levels when you mix. This will also let you EQ and compress them individually.

Other Four-Input Interface Options

Other options are also available when recording with only four inputs:

♦ Using a Blumlein Stereo Pair instead of the traditional overheads can produce great stereo imaging. Place the pair just behind and above the drummer's head, aimed forward and pointing slightly down toward the center of the kit. This provides a drummer's perspective for the recorded drums. Adding a kick and snare mic to this setup can give excellent overall results in a good sounding room.

♦ Use a small submixer to premix two or three tom microphones and a pair of stereo overhead mics down to stereo, then patch the submixer out to two of the audio interface's line-input channels. Record the premixed toms and cymbals to a stereo pair of tracks, while recording the kick and snare to their own tracks.

Six-Channel Interface with Six Mics

For this configuration, you would retain the stereo pair with kick and snare, but now you can add two mics for particular setups.

♦ If a four-piece drum kit has two toms, to emphasize toms in the recording use the two extra mics as spot mics on the toms.

♦ When recording a larger kit, aim each additional mic between pairs of toms to cover up to four toms.

♦ If a song emphasizes the hi-hat, you may want an extra mic for it. Use the sixth channel for a figure-8 or omni-directional room mic.

Eight-Channel Interface with Eight Mics

An eight-channel interface has enough inputs to cover a five-piece drum kit fully—kick, snare, and stereo overheads, plus three tom mics and your choice of a hi-hat or room mic. When recording jazz, consider using the last mic as a spot mic for the ride cymbal. Fig. 6.29 shows the kit with tom mics added, and the side overhead pulled back and up a bit, but retaining the same distance from the center of the snare as the other overhead mic.

Figure 6.29 Eight-mic setup with three tom mics. The mic to the side of the floor tom has been moved overhead and back a bit.

Ten (or More)-Channel Interfaces with Ten (or More) Mics

In addition to the basic five-piece drum kit configuration outlined for an eight-channel interface, you can add both the hi-hat mic and a stereo pair of room mics, or a mono room mic and a spot mic on the ride cymbal, or a second kick or snare mic.

Figure 6.30 shows a modified Glyn Johns overhead mic setup with two mics on the kick, one on the snare's shell side, one on each tom, and a figure-8 mic behind the drummer's head. The figure-8 mic's two lobes point to the "sides" of the room to pick up room reflections instead of the kit's direct sound. Always check any secondary mic for phase coherence with the main kick mic.

Figure 6.30 A full ten-mic setup with two overhead mics, two mics on the kick, one mic for each tom, a snare mic, hi-hats mic, and a room mic.

Individual Drum Mics and Mic Placement

If you can't afford a new multi-input audio interface and the drum mics to plug into it, don't despair: start with what you have; then augment it later. Ideally, you'll want a variety of mic types for drum kits.

Remember to factor in the cost of mic stands and cables. Good stands simplify positioning your mics during setup, and they'll be less likely to slip out of position. Also, take good care of your cables. Lay them out neatly around the kit, so they're less likely to be stepped on or tripped over (Fig. 6.31).

Figure 6.31 Proper cable routing makes damage less likely.

Miking the Kick

Large-diaphragm dynamic mics are the most popular choice for recording the kick. Typical mics include:

- ♦ Electro-Voice RE20
- ♦ Electro-Voice RE30
- ♦ Audio-Technica ATM250
- ♦ Shure Beta 52
- ♦ Audix D6
- ♦ AKG D112

Positioning

If your kick has no hole in the front head, mic it from a few inches in front of the head for a full sound. For more attack, place the mic around the other side (near the drummer's foot and the kick beater).

If the drumhead has a hole, insert the mic about three-quarters of the way into it, and aim the mic about midway between where the beater strikes and the tom side of the kick-drum shell (Fig. 6.32). For more attack, aim the mic more toward the beater. For more resonance, angle the mic more toward the shell.

Figure 6.32 Aiming the kick mic midway between where the beater hits (white circle in the left image) and the side of the shell gives a good balance of shell resonance and beater attack. Rotating the mic toward the beater adds attack; rotating toward the shell adds resonance. If the kick doesn't need to be highlighted, sometimes engineers will pull the mic back a bit to capture more of the kit sound.

Supplementing the Kick Sound

A second kick mic like the Yamaha Subkick (Fig. 6.33) can capture extra low-frequency wallop and bottom.

Figure 6.33 This closeup shows the Yamaha Subkick supplementing the main kick drum mic.

This type of mic is basically a speaker in reverse—air hits the speaker cone, and its movement generates a voltage. This setup lets you record each kick mic to its own track and blend them when mixing. Make sure any secondary mic is in phase with the main kick mic. On mixdown, nudge tracks into alignment if needed.

Miking the Snare

The usual choice for recording snare is a small-diaphragm dynamic mic, like the following:

- Shure SM57

- Audix i5

- Audio-Technica ATM650

However, both large- and small-diaphragm condenser mics can work well. You may need to engage the mic's pad switch (if present) to avoid overload. These are good condenser choices for snare:

- Audio-Technica Pro 37

- RØDE NT5

- AKG C414

Positioning

Top-miking the head is the most common approach, with the mic placed a few inches above the drum's rim and angled in so it's aimed toward either the head's center or a point two-thirds of the way toward the center. (See the illustration in Figure 6.27 earlier in this chapter.)

An alternate approach is to point the mic at the side of the snare drum (the shell) instead of the head. The earlier Figure 6.28 shows this positioning viewed from the top, while Figure 6.34 below shows side miking from a side angle. This technique can give an even sound that balances attack, resonance, and snare rattle. Avoid aiming the mic directly at the air vent on the drum's side.

Figure 6.34 The side capture option isn't always about the sound—if the drummer has questionable stick control, side-miking can save your mic by getting it out of the stick's path.

Supplementing the Snare Sound

Some engineers place a second mic a few inches under the drum, aimed at the snares themselves. To avoid phase cancellation when mixing, you'll need to invert this mic's polarity (phase) if combined with the main snare mic.

Miking Rack Toms

Dynamic mics usually get the nod for toms, and their ruggedness gives them a better chance of survival if they're subjected to an unexpected stick hit. The following are all common dynamic mic choices for toms:

- ◆ Audix D2

- ◆ Audix D4

- ◆ Sennheiser MD 421

- ◆ Sennheiser e 604

- ◆ Audio-Technica ATM25

- ◆ Audio-Technica ATM250

- ◆ Shure SM57

For drummers with good stick control who won't hit the mic, the next mics are suitable condenser types. As with the snare, you may need to engage the mic's pad switch:

- ◆ CAD M179

- ◆ AKG C414

- ◆ Neumann TLM102

- ◆ Audio-Technica Pro 37

Positioning

With either mic type, the placement approach is similar to that for snare—place the mic above the rim's edge, and angled in so that it points about midway between the rim and the middle of the drum (Fig. 6.35).

Figure 6.35 Toms being miked by a pair of Audix D2 mics. With toms, it's important to place the mics where they won't get hit by an errant stick.

Miking the Floor Tom

Large-diaphragm dynamic or condenser mics are the go-to types for floor toms. The same models recommended for rack toms will work, but the Audix D4, Sennheiser MD 421, and Audio-Technica ATM250 seem particularly well-suited to the task (Fig. 6.36).

Figure 6.36 Floor tom being miked by an Audio-Technica ATM250.

Using Overheads

Small- and medium-diaphragm condensers are common for drum overhead duties. These work well:

♦ Mojave MA-101

♦ Audio-Technica AT4041

- Oktava MK-012

- MXL 603

- AKG C451 B

- AKG C414

Ribbon mics can also work very well for drum overhead miking, especially if you're trying to tame overly bright or enthusiastically played cymbals. Here are some popular options:

- Royer R-121

- Beyerdynamic M 160

- Cascade Fat Head II

Positioning

Placement options for overhead mics are numerous. As mentioned previously, the overhead placement with the classic Glyn Johns technique places one mic over the center of the kit—usually above a spot somewhere between the rack tom, snare drum, and hi-hat, with the second overhead mic placed much lower, near the far side of the floor tom, aimed across and toward the hi-hats (Fig. 6.37).

Figure 6.37 Note the mics toward the upper left and middle right, which are the same distance from the snare drum's center.

The two overhead mics should be equidistant from the snare drum's center. A tape measure or a three- to four-foot length of twine or cable can verify that the spacings match. This ensures that the snare drum hits remain "centered" in the stereo sound field captured by the overhead mics. Adding a kick-drum mic, and possibly a snare mic, to this overhead arrangement is often all you need to capture a great drum sound with only three or four mics.

Other overhead placement options include putting a spaced pairs of microphones or an XY stereo pair directly above the kit's center. Place them about two to four feet above the top of the toms if your room has a low ceiling; move them up a bit higher in a tall room. For a drummer's perspective, place an XY pair of cardioid microphones or a Blumlein Stereo Pair (crossed at right angles) with bidirectional ribbon or condenser microphones just above, and directly behind, the drummer's head. Angle them slightly downward and aimed toward the kit's center.

Miking Hi-Hats

You'll usually capture more than enough of the hi-hats in the overheads or from bleed into the snare mic. For more hi-hat in the mix, try a small-diaphragm condenser like the following.

- SE Electronics sE2a

- AKG C451 E

Most ribbon mics can also work well here and can help subdue overly bright hi-hats. Avoid pointing toward the edge of the hats or miking them from the side, as these locations can produce air blasts. Start with the mic placed four to six inches above the hi-hats and three to four inches in from the outer edge (Fig. 6.38).

Figure 6.38 A separate hi-hat mic isn't always needed, but if you have enough channels available, it may come in handy.

Using Room Mics

A large-diaphragm vocal condenser mic can often serve double-duty as an effective room mic. Ribbon mics also work well as room mics.

General Tips and Guidelines for Drums

Following are some general tips for miking and recording drums:

- **Leave some dynamic range headroom to avoid clipping your preamps or recordings.** A good average level is around –15 to –18 dB below 0 on your interface or software's VU meters. The occasional peak or hard hit can be higher, but the red clipping indicator should never light up when recording tracks.

- **Work around the drummer as much as you can.** Drummers will be more comfortable (and probably play better) if you don't alter their setup too much. Sometimes that requires taking a different approach from conventional drum mic placements to accommodate unorthodox drum setups.

- **Take it on the road.** If your room isn't up to par, take the recording to a different drum room such as a local hall, church, or warehouse. Laptop computers and mobile interfaces make this relatively easy to do. A great-sounding room will make a significant difference in a final recording's sound.

- **Experiment with setting up the kit in different areas of the room.** Some spots will sound better than others. If you find a sweet spot, take pictures and measurements so you can duplicate the setup whenever needed.

- **Walk around the room and listen while the drummer plays.** Listen for any obvious problems (squeaky pedals, rattling hardware, out-of-tune drums), and consider how well the drummer balances the various kit elements while playing. Listen for the impact and relative balance of all the different components of the kit, and pay attention to the ratio of direct sound versus room reflections and ambience. Place your room mics wherever you hear the best balance of all these sounds.

- **Use only the mics you need.** If a drummer hits the ride cymbal really hard, the overheads might capture enough so that a spot mic on the ride isn't needed. The same can be true with hi-hats. Although a spot mic aimed at the hi-hat may be helpful, there can be enough leakage from the room mics, overhead mics, and the snare and tom mics.

- **Fix cymbal problems.** Some drummers clobber the brass while hitting the skins relatively softly. This may be because cymbals are often not well-miked in small clubs, so drummers get used to hitting them hard. In the studio, the opposite is often needed—less force on the brass, and more consistent hits on the drumheads. You can ask the drummer to try hitting the drums harder, but they'll often revert back to their muscle memory. Swapping out quieter, faster-decaying cymbals can help. Another option is to raise the cymbals' and hi-hat a few inches above where the drummer normally places them. This moves the cymbals farther away from any tom and snare mics to reduce leakage, but also changes the angle at which the drummer strikes the cymbals. This can result in less forceful hits.

♦ **While positioning your microphones, check the mixed drum sound in mono to make sure the mics are in phase.** Phase cancellation sounds hollow and lacks bass when summed to mono. Inverting the phase on a mic preamp channel may help.

♦ **Adjust mic placement as needed.** If the drummer sets the cymbals really low and isn't comfortable moving them higher, the tom mics might not fit under them. You may have to aim mics toward the tom shells or the toms' bottom heads instead.

♦ **Take these general-purpose rules as suggestions only.** For any given track or song, do what sounds good to you with a particular kit and drummer in a particular room. Mic placement suggestions are guidelines, not gospel.

♦ **Avoid setting up in the center of the room.** Acoustic phase cancellations are more likely to occur in the center than at other spots within the room. Also be careful of the bass buildup that happens in corners. This can give tracks an undefined, muddy character.

♦ **Talk to the drummer, Part 1.** If they've had prior recording experience, they may have insights about what has—and hasn't—worked in the past.

♦ **Talk to the drummer, Part 2.** Make sure your mics are placed out of the way of their playing, so a stick doesn't hit them.

♦ **Talk to the drummer, Part 3.** Ask if they *won't* be using any elements of the kit. There's no need to worry about how the kit-mounted cowbell will sound in your mix if it won't get played.

♦ **Try applying small changes; they can make a big difference.** Before you abandon a mic setup for something completely new, try adjusting what you already have. For example, rotating the kick mic an inch or two more toward where the beater strikes (or away from it and towards the drum shell) can make a big difference in the amount of attack on each kick hit. Moving the mic a few inches farther into the drum or farther back from the beater can also change the sound.

♦ **Remember that distance equals depth.** Close mics tend to capture an "in your face" sound, while placing mics farther back captures more of the room so the drums sound farther away from the listener. Bleed will increase when mics are farther away, but there will be less bass boost if the proximity effect is in play.

♦ **Pay attention to where the back of the mic points—That can be just as important as where you've aimed the front.** You'll aim the mic toward the sweet spot of what you're recording, but if you can aim the mic's null point toward what you want to reject, that's even better. For example if you're picking up too much hi-hat, try to aim the snare mic's dead spot toward the hats.

♦ **Whenever possible, avoid having a nearby mic on the same plane as a cymbal, or aiming at the thin edge of a cymbal.** This gives an unnatural, phasey sound as the cymbal pivots in and out of position relative to the mic.

Horns and Saxophone

Let's look at miking the three instruments found most often in a small horn section—trumpet, trombone, and saxophone.

Trumpet and Trombone

Trumpet and trombone are brass instruments, with a mouthpiece and a bell at opposite ends of a long metal tube. Players produce notes by vibrating their lips into the mouthpiece at one end of the tube. The sound travels down the tube and out the other end at the bell. The player can adjust the tube's length and the notes the instrument makes by varying the pitch produced by the vibrating lips, and in the case of the trumpet, also by adjusting three valves that route the sound through different lengths of brass tubing. There are valve trombones too, but most trombones have a slide mechanism, where moving the slide in or out decreases or increases the tube's length.

Trumpets and trombones are *loud,* and both instruments require a mic that can handle very high SPL sound sources. When using a condenser mic, you'll usually need to engage its pad switch to keep it from overloading. The DPA d:vote 4099 series of condenser instrument mics come in a low-sensitivity version optimized for high-level sound sources. These mics are highly recommended for both trumpet and trombone.

For dynamic mics, popular choices include the following:

♦ Electro-Voice RE20

♦ Audix D4

♦ Sennheiser MD 441

Ribbon microphones such as the following are also effective for brass instruments:

♦ Cascade Fat Head II

♦ Coles 4038

♦ Royer R-121

♦ Royer R-101

For a brasher, brassier, more piercing sound, try a condenser mic. Reach for a dynamic or ribbon mic to tame some high end.

Trumpet and trombone are fairly directional instruments. If you place the microphone in front of the bell (Fig. 6.39), you'll hear the brightest, most detailed sound. Moving off-axis toward the side of the player gives a more muted, less direct sound.

Figure 6.39 A DPA d:vote 4099 miking a trumpet for the brightest sound.

You can use the placement to your benefit if a trumpet or trombone is too bright—move the mic off-axis a bit instead of reaching for EQ. Miking on-axis and directly in front of the bell at a distance of one to three feet is the recommended starting position. An alternative approach is miking the reverse side of the bell (from the player's side) at a distance of about 6 to 12 inches. This gives a mellower sound than miking the front of the bell.

Saxophone

While the saxophone isn't technically a horn, but rather a woodwind instrument, it's still a vital component of most brass sections, and many horn parts on recordings use saxophones.

Saxophones come in four common sizes and note ranges: the small, high-pitched and often straight-bodied soprano, the more familiar tones and curves of the tenor and alto sax models, and the large, beefy baritone filling out the low end. Like the trumpet and trombone, all saxophones have a mouthpiece at one end and a large bell at the other end. Although you might think that's where the sound comes from, a sax produces sound when the player blows and vibrates a reed mounted to the mouthpiece.

A saxophone also has numerous holes along the length of its body, with pads that cover the holes. Adjusting the keys that open and close these pads changes the pitch and the location from where the sound emanates.

As a result, a mic placed to point directly at and deep down into the sax's bell usually fails to offer a balanced sound. However, don't discount that mic position entirely—it's so commonly used that its unnatural sound is one that many people are familiar with, and it is sometimes used intentionally for effect.

For a fully balanced sound, place the mic where it captures not only the sound of the resonating bell, but also the sound coming out of the various holes along the length of the instrument (Fig. 6.40).

Figure 6.40 A DPA d:vote 4099 is angled somewhat away from the bell, and points at the body itself for a more balanced sound.

Choosing a Saxophone Mic

Small- or large-diaphragm condenser mics usually deliver the brightest, most detailed sound and capture all the rasp, spit, and sizzle the instrument can create. For a somewhat rounder and warmer sound, try a ribbon mic. The beyerdynamic M 160 is good for this application because it has a fairly extended high-frequency response by ribbon mic standards, yet still has a warm sound that complements the instrument's sound exceptionally well. Darker-sounding ribbon mics can help de-emphasize excessive brightness or stridency. Some people also like dynamic mics (e.g., Sennheiser MD 441 and MD 421), which can work well on sax. DPA's d:vote 4099 is particularly well-suited for players who like to move around. Because the 4099 attaches to the instrument, it's largely immune to even the wildest gyrations from enthusiastic players.

Try aiming the mic directly at the middle of the player's two hands, and approximately in the middle of the sax. Except with a straight-bodied soprano sax, this mic position points the mic in the general direction of the bell, but also captures the sound coming from the various pads.

If the room sounds good, explore mic positions that are slightly farther back and away from the instrument. With a standing player, place a mic about four to six feet away and three to five feet above the floor, with the mic angled so it aims at a spot directly in between the player's two hands. This will usually give a well-balanced sound that captures not only the entire instrument, but also a bit of room ambience. For more ambience, move the mic back farther; for less, move it in closer. Also consider different mic polar patterns—a figure-8 gives less ambience than an omni-directional pattern, and a cardioid pattern gives less room ambience than a figure-8 polar pattern.

Key Takeaways

♦ Encourage singers to use good mic technique—move farther away when singing louder, and closer for softer sections.

♦ A pop filter can not only reduce pops, but provide a safety zone between the mic and the singer.

♦ Having background singers farther away from their mic(s) gives more depth, which can provide a good contrast with the lead vocal.

♦ For guitar amps, small-diaphragm mics are the go-to choice—but condensers can give a brighter sound, while ribbon mics can accentuate note attacks.

♦ The sound varies considerably as you move a mic from the center of a speaker toward the edge.

♦ You don't need a super-loud amp or lots of distortion in the studio; dialing these back can improve the recorded sound.

♦ It's not always best to close-mic an amp. Extra distance can capture more of the room sound.

♦ Small-diaphragm condenser mics are popular for acoustic guitar.

♦ Using one mic can eliminate phase issues with acoustic guitar, but requires more care in placement. Adding a second mic gives stereo, and XY miking can also work.

♦ Miking rotary-speaker cabinets can be a challenge because of the mechanical noises they make. Recording with one mic is possible, but having separate mics for the horn and drum is better.

♦ Bass amps use the same general techniques as guitar amps. However, another signal, like the plain bass sound recorded direct, often complements the miked sound.

♦ If you record bass direct and miked, be aware that phase differences may result that require bringing the tracks into alignment with your host software.

♦ A drum kit isn't an instrument, but a collection of different instruments—kick, snare, toms, cymbals, and hi-hat.

♦ A limited number of audio interface inputs or mics can make drum miking more challenging.

♦ It's possible to record a solid, defined drum sound with only four mics.

♦ Overhead mics are important to capture the drum kit's overall sound.

♦ Large-diaphragm dynamic mics are preferred for kick drums. However, there are also dedicated kick-drum mics, and sometimes both are used together.

♦ Small-diaphragm dynamic mics are the usual choice for snare, but large- and small-diaphragm condenser mics can give a crisper sound if desired.

♦ Dynamic mics are the main choice for toms. Condenser mics sometimes get the nod if you're confident the drummer won't hit them accidentally.

♦ Large-diaphragm mics, either condensers or dynamics, are the go-to mics for floor toms.

♦ There's usually enough bleed from the hi-hat into other mics that a separate hi-hat mic isn't needed, but if it is, try adding a small-diaphragm condenser mic.

♦ Trumpet and trombone are very loud. Make sure your mic can handle the level.

♦ The common mic position for brass is pointing into the instrument's bell.

♦ Saxophone is a woodwind instrument. The most accurate-sounding mic position doesn't point down into the bell, but aims across the bell so that it captures sounds made by the instrument itself.

Chapter 7

Additional Tips

Some concepts don't fit neatly into chapters, so here's a grab bag of microphone-related tips.

Miking Synthesizers and Electronic Instruments

No, this isn't a joke. While synths and drum machines don't *need* miking, sometimes they can have an unnatural sound when their direct signals are mixed with acoustic or amplified instruments. Getting a mic into the picture is often the solution.

With electronic instruments, space truly is the final frontier. Although we listen through air, hardware and virtual synths generate electrical signals that need never reach air until we hear the final mix. Compared to acoustic instruments, synth sounds are relatively static—especially since the rise of sample-playback machines. Yet our ears are accustomed to hearing evolving, complex acoustical waveforms that are unlike synth waveforms, so creating a simple acoustic environment for the synth is one way to end up with a more interesting, complex sound. This technique can also help synths blend in with tracks that include miked instruments, because they usually include some amount of room ambience (even with fairly "dead" rooms).

Sending some (or even all) of your synth's signal to an amp and miking it can give that feeling of air. For a clean sound, the Fender Twin is an old standby, but an amp like Line 6's DT25 or DT50 is an excellent choice. These models make it possible to change the amp's topology in the analog domain to generate anything from clean to dirty sounds.

Virtual instruments can take advantage of this technique too—just pretend you're re-amping a guitar track. Send the virtual instrument output directly to a hardware audio output on your computer's audio interface (this assumes that your interface has multiple outputs), pad down the output, run it into your amp, then feed the mic into a spare audio interface input and record this signal to your DAW. There will likely be some delay introduced when going from digital to analog then back to digital again, but you can always compensate for this by "nudging" the miked track a little bit earlier.

Another way to add the feel of an acoustic space to a synth is to record the sound of the keys being hit. (Sometimes a contact mic works best.) Mix this very subtly in the background—just enough to give a low-level aural cue. You may be surprised at how much of a natural quality this technique can add to synthesized keyboards.

Miking a Band

Miking a whole band can be a bit daunting, but it's not nearly as difficult if you break the process down into individual elements. Start by miking the band's foundation—bass and drums, and then continue with vocals, acoustic guitar, and electric guitar, along with synths and electronic instruments.

Whether you record each part separately or the entire band at once is up to you. For many years, it was *de rigueur* to record everything separately to a click track, but for musical and musician-interaction reasons, recording the rhythm section (drums, bass, rhythm guitar, etc.) together whenever possible provides the most cohesive feel. Then you can overdub the lead parts and vocals. Although recording multiple musicians together requires more input channels on your audio interface and possibly some baffling to reduce leakage, it's usually a more natural process for the musicians, and allows them to play off one another.

The Pros and Cons of Close-Miking

While close-miking techniques can reduce the influence of room acoustics, the location of your instruments and mics in a room can affect the overall sound quality dramatically. Avoid boomy-sounding corners, and don't set up too close to walls that might contribute unnatural reflections. If your room is less than ideal, move the mics in close and use as much baffling and broadband absorption around your instruments and mics as possible. You'll have to add in early reflections, ambience, and reverb at mixdown, but this is often a better approach than capturing negatively-colored room reflections that you won't be able to remove from your tracks later.

Also, don't close-mic everything if you can avoid it. If you do, then everything may wind up fighting for the up-front placement in the mix. Give some advance thought to varying the mic distances for various sound sources—consider placing the mics farther away from for supportive elements and closer for main elements. This configuration helps the parts sit in their own space.

If you want to capture more room ambience but still close-mic a sound source, consider using omni mics. They can capture more of the room's character and ambience than cardioid mics, but they won't have any proximity effect. In the event that it works well, you're covered. If not, boosting the bass during mixdown gives pretty much the same effect.

Get the Sound Right at the Mic

EQ and compression are fantastic tools, but don't expect to use them in the mix. Consider changing an amp's tone controls, adjusting the mic position, or even using a different mic. If the ribbon mics you're using for the drum overheads sound too dark, try some condenser mics instead; if the guitar amp sounds too bright, re-position the mic. Sounds that don't need fixing always seem to sound better for some reason, even when you can add processors to get the kind of sound you like.

Key Takeaways

◆ Although most people record synthesizers and drums direct, feeding them through amps and miking them can give more interesting textures that also blend better with other miked instruments.

◆ Recording the sound of fingers hitting keyboard keys, and mixing it in at a low level in the background, can give a more authentic feel.

◆ Miking an entire band can be daunting, but dealing with it one piece at a time makes the process easier—do bass and drums first, then vocals, guitars, keyboards, and other instruments.

◆ There's a certain vibe that can be captured by recording musicians playing together, although managing leakage will be an issue.

◆ Close-miking makes room acoustics less relevant.

◆ If the acoustics are questionable, try adding electronic ambience (e.g., reverb and delay) to the close-miked sound.

◆ Use omni mics if you want to do close-miking and still capture a significant amount of room ambience.

◆ Choosing the right mic from the start can give better results than reaching for electronic processing like EQ and dynamics to shape the sound.

About the Authors

Phil O'Keefe is a multi-instrumentalist, recording engineer/producer, and the Senior Editor of Harmony Central (www.harmonycentral.com). He served as a staff producer/engineer for Blonde Vinyl records, where he worked on albums by Michael Knott, LSU, Breakfast With Amy, and Fiction Alley. As the owner and chief engineer of Sound Sanctuary, he has worked with artists such as Ska/Punk legends Voodoo Glow Skulls, multiplatinum rockers Alien Ant Farm, Jazz vocal phenomenon Jules Day, singer-songwriter Zak Claxton, Rex Smith, Christian rocker Jeff Fenholt and Dio guitarist Craig Goldy, and existential folk-rock artist John McGill, among many others.

As a music journalist, Phil's articles and reviews have appeared in Electronic Musician, Guitar Player, Keyboard, and other publications, and he is a former featured monthly columnist for EQ magazine. He has been involved online since the very beginning of the world wide web, having co-moderated an early music-centric FIDOnet BBS. Before moving his popular Studio Trenches forum to Harmony Central in 2005, he served as a moderator on the Musicplayer forums alongside such industry heavyweights as Roger Nichols, Chris Stone, Ed Cherney, George Massenberg, Bill Dooley, and Craig Anderton.

Musician/author **Craig Anderton** is an internationally recognized authority on music and technology. His onstage career spans from the 60s with the group Mandrake, through the early 2000s with electronic groups Air Liquide and Rei$$dorf Force, to the "power duo" EV2 with Public Enemy's Brian Hardgroove, and EDM-oriented solo performances.

He has played on, produced, or mastered over 20 major label recordings, did pop music session work in New York in the 1970s on guitar and keyboards, played Carnegie Hall, and more recently, has mastered well over a hundred tracks for various artists.

In the mid-80s, Craig co-founded *Electronic Musician* magazine. As an author, he's written over 26 books on musical electronics and over a thousand articles for magazines like *Keyboard, Sound on Sound, Rolling Stone, Pro Sound News, Guitar Player, Mix,* and several European publications.

Craig has lectured on technology and the arts (in 10 countries, 38 U.S. states, and three languages), and done sound design work for companies like Alesis, Gibson, Peavey, PreSonus, Roland, and Steinberg.

Please check out some of his music at youtube.com/thecraiganderton, visit his web site at craiganderton.com, and follow him on twitter @craig_anderton.

www.ingramcontent.com/pod-product-compliance
Ingram Content Group UK Ltd.
Pitfield, Milton Keynes, MK11 3LW, UK
UKHW051925290426
470527UK00005B/42